Graph Theory in America

Kenneth Appel and Wolfgang Haken contemplate their solution to the four color problem.

Graph Theory in America

The First Hundred Years

ROBIN WILSON
JOHN J. WATKINS
DAVID J. PARKS

PRINCETON UNIVERSITY PRESS
PRINCETON AND OXFORD

Published by Princeton University Press
41 William Street, Princeton, New Jersey 08540
99 Banbury Road, Oxford OX2 6JX

press.princeton.edu

ISBN 9780691194028
ISBN (e-book) 9780691240657

Library of Congress Control Number: 2022943051

British Library Cataloging-in-Publication Data is available

Editorial: Diana Gillooly, Kristen Hop, and Kiran Pandey
Production Editorial: Karen Carter
Text Design: Carmina Alvarez
Jacket/Cover Design: Jessica Massabrook
Production: Jacqueline Poirier
Publicity: Carmen Jimenez and Matthew Taylor
Copyeditor: Nancy Marcello

Jacket/Cover Credit: Jacket images clockwise: Chemical diagram by J. J. Sylvester from the *American Journal of Mathematics* 1 (1878), 64–125, plate I. Used with permission by the Department of Mathematics, University of Illinois Urbana-Champaign. James Joseph Sylvester. Stipple engraving by G. J. Stodart / Wikimedia Commons. Courtesy of Wellcome Images. Sylvester's Graph. Used with permission by the Department of Mathematics, University of Illinois Urbana-Champaign. Reducible configurations by Appel and Hagen. Kenneth Appel (*top*) and Wolfgang Haken (*bottom*). Used with permission by the Department of Mathematics, University of Illinois Urbana-Champaign.

This book has been composed in Baskerville 10 Pro with Gill Sans Std.

Printed on acid-free paper. ∞

Printed in the United States of America

10 9 8 7 6 5 4 3 2 1

CONTENTS

FOREWORD

Gary Chartrand

How often is it, I wonder, that when young students are introduced to mathematics subjects such as algebra, geometry, and perhaps even calculus, they ask: "When and where did these subjects come about, and who created them?" As these students become older, attend college, and encounter mathematical subjects with more sophisticated-sounding names, such as abstract algebra, analysis, topology, and combinatorics, the same questions apply.

My guess is that questions such as these are rarely asked, if at all. One reason may be that people tend to think that mathematics has been around forever, that there's nothing new to discover, and that mathematics is not due to any particular individuals or did not begin in any particular place. While there is no single answer to these questions, one response is clear. These subjects did not originate in America. In some instances, there was no United States as yet—or, if so, it was in its infancy. Several mathematical subjects grew out of people, primarily from Europe, but also from Asia and Africa. These people were not always mathematicians—but they were people with an abundance of mathematical curiosity.

This book deals with one area of mathematics in particular—my all-time personal favorite: graph theory. A youngster among the fields of mathematics, it is fascinating for a wide variety of reasons. It has connections with many other subjects, both inside and outside mathematics—some were highly scholarly, while others grew out of interesting and imaginative problems that were just plain fun.

The questions and problems that have arisen in this area are often easy to state, but notoriously difficult to answer or solve. Nevertheless, we, the readers, find ourselves in the enviable position of being about to be led through the story of how much of this subject found its way to North America and was impacted by Americans and Canadians—told by a master mathematical explorer, historian, and storyteller, Robin Wilson, along with his fellow authors John Watkins and David Parks. These exceptional individuals have enjoyed a lifelong interest in this exciting area of mathematics and in the history behind it.

The early topics in graph theory were often passed along from one individual to another by means of personal conversations or handwritten letters, as no other means of communication were accessible during the time. Later, the communication of topics became available through public lectures and articles in magazines, while later still, books were written that collected, summarized, and solidified the important information known at that time. These efforts were followed by travel to international conferences, which made graph theory a worldwide subject. Finally, email communication, along with the Internet, has opened up the subject to everyone.

So, sit back, get comfortable, with this book in hand or a computer screen before you, and be prepared to enjoy a story of how one remarkable area of mathematics, with humble beginnings deeply rooted in Europe, traveled to and through North America, and how Americans and Canadians came to influence this journey.

Gary Chartrand
Western Michigan University

PREFACE

Over the hundred years from 1876 to 1976, graph theory underwent a fundamental transformation, both worldwide and in North America. In 1876, the Englishman James Joseph Sylvester took up his appointment as the first professor of mathematics at the newly founded and research-oriented Johns Hopkins University in Baltimore, Maryland, where his inaugural lecture outlined the connections he saw between graph theory and chemistry. It was shortly after this that he introduced the word *graph* in our modern sense. In 1976, after much activity by many people, the four color problem was finally solved by Kenneth Appel and Wolfgang Haken of the University of Illinois.

This book is based on a doctoral dissertation by David Parks for the Open University, UK, under the supervision of Robin Wilson, and chronicles the historical development of graph theory in America between these two significant years. Whereas many of the featured mathematicians spent their entire career working on problems in graph theory, a few (such as Hassler Whitney) began their career in this subject but later became better known for their researches elsewhere, while others (such as C. S. Peirce, Oswald Veblen, and George Birkhoff) made excursions into graph theory while continuing their mainstream work in other fields.

Interwoven between the main chapters of this book are "interludes" that help to set the scene for the main narrative. We open with a description of how American universities developed up to the 19th century, while the two later interludes present the parallel development of graph theory in Europe. We conclude with an "Aftermath" that describes further achievements up to 1976 and mentions some important developments after this.

A special feature of this book is the inclusion of short summaries of some specific publications that influenced the subject's development. Listed after this Preface, these range from Birkhoff's groundbreaking paper on the reducibility of maps, and Whitney's seminal writings on the planarity of graphs, to the fundamental research of Gerhard Ringel and J.W.T. Youngs into the drawing of graphs on topological surfaces.

These summaries can be omitted by those not interested in the mathematical details.

No prior knowledge of graph theory is required when reading this book, which aims to explain the historical development of the subject in simple terms to a general reader interested in mathematics. However, the mathematical level inevitably varies somewhat, especially in the later chapters, and readers who are interested mainly in the historical narrative, and in the personalities involved, can safely pass over any technical material. The book concludes with a glossary of graph theory terms used throughout the book.

The authors wish to thank the Open University and the Colorado College for encouragement and support. We should also like to thank Gary Chartrand, Karen Hunger Parshall, Bjarne Toft, and Matjaž Krnc for their helpful comments on the manuscript, and Vickie Kearn, Susannah Shoemaker, Lauren Bucca, Kristen Hop, Diana Gillooly, Karen Carter, and Eric Crahan of Princeton University Press for all their help in bringing this book into being.

Robin Wilson, John J. Watkins, and David J. Parks

FEATURED PAPERS

The publications that are summarized in the text are as follows:

Chapter 1

J. J. Sylvester (1878): "On an application of the new atomic theory to the graphical representation of the invariants and covariants of binary quantics,—with three appendices"
A. B. Kempe (1879): "On the geographical problem of the four colours"
W. E. Story (1879): "Note on the preceding paper"

Chapter 2

O. Veblen (1912): "An application of modular equations in analysis situs"
George D. Birkhoff (1912): "A determinant formula for the number of ways of coloring a map"
George D. Birkhoff (1913): "The reducibility of maps"

Chapter 3

Philip Franklin (1922): "The four color problem"
J. Howard Redfield (1927): "The theory of group-reduced distributions"

Chapter 4

Hassler Whitney (1932): "A logical expansion in mathematics"
Hassler Whitney (1932): "Non-separable and planar graphs"
Hassler Whitney (1935): "On the abstract properties of linear dependence"

Chapter 5

G. D. Birkhoff and D. C. Lewis (1946): "Chromatic polynomials"
L. R. Ford Jr. and D. R. Fulkerson (1956): "Maximal flow through a network"
R. C. Prim (1957): "Shortest connection networks and some generalizations"
W. T. Tutte (1959): "Matroids and graphs"

Chapter 6

Oystein Ore (1960): "Note on Hamilton circuits"
Gerhard Ringel and J.W.T. Youngs (1968): "Solution of the Heawood map-coloring problem"
K. Appel and W. Haken (1976): "Every planar map is four colorable"

CHRONOLOGY OF EVENTS

In this chronology we list the most important publications and events that are featured in this volume.

1636	Harvard College (later, Harvard University) is founded in Cambridge, Massachusetts.
1701	The Collegiate School (later, Yale University) is founded near New Haven, Connecticut.
1735	Leonhard Euler solves the Königsberg bridges problem.
1746	The College of New Jersey (later, Princeton University) is founded in New Jersey.
1750	Euler states his "polyhedron formula", $F+V=E+2$.
1812–13	Simon-Antoine-Jean L'Huilier extends Euler's formula to orientable surfaces.
1827	King's College (later, the University of Toronto) is founded.
1847	Gustav R. Kirchhoff writes on electrical networks and introduces spanning trees.
1852	Francis Guthrie poses the four color problem for maps.
1856	Thomas P. Kirkman and William R. Hamilton investigate "Hamiltonian cycles" on polyhedra.
1857	Arthur Cayley writes his first paper on trees.
1861	The Massachusetts Institute of Technology (MIT) is founded in Boston.
	The first doctoral degrees are awarded, at Yale College.
	Johann B. Listing discusses "spatial complexes" in topology.
1862	The Morrill Act is passed, allowing for expansion in higher education.
1875	Cayley enumerates certain chemical molecules.

Now begin "The First Hundred Years" that are the focus of this book

1876 Johns Hopkins University is founded in Baltimore, Maryland, and James Joseph Sylvester is appointed the first professor of mathematics.

1878 The *American Journal of Mathematics* is launched.

Arthur Cayley introduces "Cayley color graphs" and revives the four color problem at a meeting of the London Mathematical Society.

James Joseph Sylvester writes on the "new atomic theory" and introduces the word "graph".

1879 Cayley shows that the four color problem can be restricted to cubic maps.

Alfred B. Kempe proposes a proof of the four color theorem in the *American Journal of Mathematics*, and William E. Story comments on Kempe's paper.

1880 Peter Guthrie Tait reformulates the four color problem in terms of coloring the boundaries of a cubic map.

1884 Tait conjectures that every cubic polyhedron has a Hamiltonian cycle.

1888 The New York Mathematical Society is founded.

1889 Cayley presents his n^{n-2} result on the number of labeled trees.

1890 The University of Chicago is founded.

Percy J. Heawood points out the error in Kempe's proof of the four color theorem, proves the five color theorem, and discusses the coloring of maps on orientable surfaces.

1891 Lothar Heffter investigates the coloring of maps on orientable surfaces, and points out the omission in Heawood's paper, leading to the "Heawood conjecture".

Julius Petersen discusses the factorization of regular graphs.

1894 The New York Mathematical Society is renamed "The American Mathematical Society".

1895 Gaston Tarry presents a method for tracing a maze.

1898 Heawood explores the congruences arising from coloring cubic maps.

Petersen introduces the "Petersen graph".

1904 Paul Wernicke introduces a new unavoidable set for the coloring of maps.

c.1905 Hermann Minkowski attempts unsuccessfully to prove the four color theorem.

1910 Heinrich Tietze investigates the coloring of maps on non-orientable surfaces.

1912 George D. Birkhoff introduces chromatic polynomials.

Oswald Veblen writes on the algebra of graphs and recasts the four color problem in the context of projective geometry.

1913 Birkhoff discusses reducible configurations in maps and introduces the "Birkhoff diamond".

1914–18 World War I (Canada enters in 1914; United States enters in 1917).

1916 Dénes König investigates 1-factors and edge colorings in bipartite graphs.

Veblen presents the American Mathematical Society Colloquium Lectures on *Analysis Situs* (published in 1922).

1920s Alfred Errera writes several papers on map coloring.

1922 Philip Franklin writes on the four color problem and proves that all maps with up to 25 countries can be colored with four colors.

1923 H. Roy Brahana publishes an important result on the topology of surfaces.

1926 Otakar Borůvka solves the minimum connector problem on minimum-length spanning trees.

Clarence N. Reynolds Jr. proves that all maps with up to 27 countries can be colored with four colors.

André Sainte-Laguë writes *Les Réseaux* (*ou graphes*) (English version 2021).

1927 Karl Menger investigates the connectedness of graphs.

J. Howard Redfield investigates enumeration problems, with applications to graphs.

1930 The Institute for Advanced Study is established at Princeton, New Jersey.

Vojtěch Jarník solves the minimum connector problem.

Casimir Kuratowski obtains a forbidden subgraphs characterization of planar graphs.

Frank P. Ramsey writes on set theory, laying the groundwork for what would become "Ramsey graph theory".

1930s The Great Depression.

1931 D. König and E. Egerváry write on matchings in bipartite graphs.

1931–35 Hassler Whitney writes seminal papers on planar graphs and duality, coloring graphs, connectivity, matroids, and other topics.

1934 Franklin proves that all maps on a Klein bottle can be colored with six colors.

1935 Philip Hall proves "Hall's theorem" on representatives of subsets (matchings).

Heinrich Heesch becomes interested in the four color problem.

Isidore N. Kagno investigates graphs on surfaces.

1936 König publishes *Theorie der endlichen und unendlichen Graphen*, the "first book on graph theory" (English version 1990).

1937 Saunders Mac Lane obtains new conditions for a graph to be planar.

George Pólya writes a groundbreaking paper on graph enumeration.

1938 Roberto Frucht proves that every abstract group is the automorphism group of some graph.

1939–45 World War II (Canada enters in 1939; United States enters in 1941).

1940 *Mathematical Reviews* is launched.

Leonard Brooks, Cedric Smith, Arthur Stone, and William T. Tutte solve the problem of "squaring the square".

Henri Lebesgue finds new unavoidable sets for map coloring.

C. E. Winn proves that all maps with up to 35 countries can be colored with four colors.

1941 Brooks presents an upper bound on the chromatic number of a graph.

Frank L. Hitchcock writes on the transportation problem.

1943 Hugo Hadwiger poses a conjecture on *k*-colorable graphs.

1946 G. D. Birkhoff and D. C. Lewis publish a major paper on chromatic polynomials.

I. N. Kagno writes on graphs and their groups.

Tutte disproves Tait's conjecture on Hamiltonian cycles in cubic polyhedra.

1947 Arthur Bernhart classifies configurations surrounded by 6-rings.

George B. Dantzig invents the simplex algorithm for solving linear programming problems.

Paul Erdős obtains bounds for a problem in Ramsey graph theory.

Tutte finds a condition for a graph to have a 1-factor and introduces the "Tutte polynomial" of a graph.

1948 Heesch lectures in Kiel on the search for an unavoidable set of reducible configurations for the four color problem, with Wolfgang Haken in the audience.

Richard R. Otter writes on the enumeration of trees.

Tutte writes his doctoral thesis on an algebraic theory of graphs.

1949 Claude E. Shannon publishes an upper bound for coloring the edges of a multigraph.

1952 Gabriel Dirac presents a sufficient condition for a graph to be Hamiltonian.

Gerhard Ringel proves the Heawood conjecture for maps on non-orientable surfaces.

Tutte presents a condition for a graph to have an *r*-factor.

1953 The William Lowell Putnam Mathematical Competition includes the "six people at a party" problem.

1954 George Dantzig, Ray Fulkerson, and Selmer Johnson write on the traveling salesman problem.

Frank Harary writes on signed graphs.

1955 Robert E. Greenwood and Andrew M. Gleason write on Ramsey graph theory.

Harary enumerates various types of graphs.

Harold W. Kuhn writes on the "Hungarian method" for the assignment problem.

1956 Lester R. Ford Jr. and D. R. Fulkerson write on the maximal flow through a capacitated network.

Joseph B. Kruskal Jr. writes on the minimum spanning tree problem.

1957 The critical path method is developed for the scheduling of tasks in a project.

Edward F. Moore introduces breadth-first search.

Robert C. Prim writes on the minimum spanning tree problem.

1958 Claude Berge writes *Théorie des Graphes et ses Applications* (English version, 1962).

1959 The first international meeting on graph theory takes place in Dobogókő, Hungary.

The University of Waterloo is founded in Ontario, Canada (originally in 1956 as Waterloo College Associate Faculties).

Edsger W. Dijkstra presents an algorithm for solving the shortest path problem.

Ringel writes *Färbungsprobleme auf Flächen und Graphen* on the coloring of graphs on surfaces.

Tutte writes on matroids and graphs, characterizing graphic and cographic matroids.

1960 Oystein Ore presents a sufficient condition for a graph to be Hamiltonian.

Alan J. Hoffman and Robert R. Singleton investigate Moore graphs.

1961 Berge introduces perfect graphs and poses the perfect graph conjecture.

P. Erdős and Alfred Rényi write on the evolution of random graphs.

1962 S. L. Hakimi writes a paper on degree sequences of graphs.

Ford and Fulkerson publish their book *Flows in Networks*.

Ore writes *Theory of Graphs*.

1963 W. Gustin introduces current graphs for solving graph embedding problems.

Ore writes *Graphs and Their Uses* for high school students.

1964 John W. Moon investigates the groups associated with tournaments.

1964–65 Vadim G. Vizing publishes fundamental results on the edge-coloring of graphs.

1965 Robert G. Busacker and Thomas L. Saaty write *Finite Graphs and Networks: An Introduction with Applications*.

Jack Edmonds writes influential papers on matchings and polynomial algorithms.

F. Harary, R. Z. Norman, and D. Cartwright write *Structural Models: An Introduction to the Theory of Directed Graphs*.

1966 The *Journal of Combinatorial Theory* is launched.

1967 The Department of Combinatorics and Optimization is founded at the University of Waterloo, Canada.

The first Chapel Hill (North Carolina) conference on combinatorial mathematics takes place.

Edmonds conjectures that there is no polynomial algorithm for solving the traveling salesman problem, and thus that $P \neq NP$.

Harary edits *A Seminar on Graph Theory* and *Graph Theory and Theoretical Physics*.

Ore writes *The Four-Color Problem*.

1968 The first Kalamazoo (Michigan) quadrennial conference on graph theory takes place.

Lowell W. Beineke presents a forbidden subgraphs characterization of line graphs.

John Moon writes *Topics on Tournaments*.

G. Ringel and J.W.T. Youngs solve the Heawood conjecture for maps on orientable surfaces.

1969 Harary writes *Graph Theory*.

Heesch writes *Untersuchungen zum Vierfarbenproblem* on the four color problem.

1970 The first Southeastern International Conference on Combinatorics, Graph Theory, and Computing takes place in Baton Rouge (Louisiana).

Moon writes *Counting Labelled Trees*.

Tutte presents his "golden identity" for chromatic polynomials.

1971 *Discrete Mathematics* is launched.

Mehdi Behzad and Gary Chartrand write *Introduction to the Theory of Graphs*.

Adrian Bondy writes on pancyclic graphs.

Stephen A. Cook writes a classic paper on complexity, introducing NP-completeness.

Heesch proposes three obstacles to reducibility for the four color problem.

1971–72 Ronald L. Graham and Henry O. Pollak prove a theorem related to telephone switching theory on partitioning complete graphs into complete bipartite graphs.

1972 Kenneth Appel begins his collaboration with Haken on the four color problem.

Richard M. Karp introduces the symbols P and NP and surveys the complexity of graph problems.

Lászlό Lovász proves the weak perfect graph conjecture.

Whitney and Tutte write on the four color problem.

Robin J. Wilson writes *Introduction to Graph Theory*.

1973 F. Harary and E. M. Palmer write *Graphical Enumeration*.

1974 Jonathan L. Gross writes on voltage graphs.

John Hopcroft and Robert Tarjan present a linear-time algorithm for testing the planarity of graphs.

John Koch joins K. Appel and W. Haken to work on the four color problem.

Ringel writes *Map Color Theorem* on the solution to the Heawood conjecture.

1976 J. A. Bondy and U.S.R. Murty write *Graph Theory with Applications*.

N. L. Biggs, E. K. Lloyd, and R. J. Wilson write *Graph Theory 1736–1936* on the history of graph theory.

Kenneth Appel and Wolfgang Haken prove the four color theorem.

This celebrated theorem brings us to the end of "The First Hundred Years"

1977 The *Journal of Graph Theory* is launched.

K. Appel, W. Haken, and J. Koch publish their proof of the four color theorem.

1978 Henry Glover and John Huneke present the forbidden subgraphs for graphs embedded on a projective plane.

Endre Szemerédi investigates random graphs, in work that contributes to his 2012 Abel Prize.

1979 Michael Garey and David Johnson write *Computers and Intractability: A Guide to the Theory of NP-Completeness*.

1983 Neil Robertson and Paul D. Seymour launch their 20-year "Graph Minors Project", during which they prove several fundamental results in graph theory.

1984 Robertson and Seymour obtain a generalization of Kuratowski's theorem for all surfaces.

1990 The Institute of Combinatorics and its Applications is founded.

1993 Robertson and Seymour with Robin Thomas prove Hadwiger's conjecture for 6-colorable graphs.

c.1994 Robertson and Seymour with Daniel Sanders and Robin Thomas obtain a revised proof of the four color theorem.

2006 Robertson and Seymour with Maria Chudnovsky and Robin Thomas publish their proof of the strong perfect graph conjecture.

2012 Endre Szemerédi is awarded the Abel Prize for his work, which included various contributions to discrete mathematics and graph theory.

2021 László Lovász and Avi Wigderson are jointly awarded the Abel Prize for their work, which included many contributions to graph theory and its applications.

Graph Theory in America

Setting the Scene
Early American Mathematics

Just sixteen years after the *Mayflower* landed at Plymouth Rock on the east coast of America in 1620, pioneers of North American colonization founded the first establishment of higher learning at Cambridge in Massachusetts. It would be more than fifty years before the second American college opened. The mission of these early colleges was mainly to transmit known knowledge at an elementary level, and it was not until the middle of the 19th century that efforts were made to provide opportunities in graduate education and to initiate research in universities and colleges.

In these opening pages we outline the early history of the American colleges, with particular attention to Harvard, Yale, Princeton, the Massachusetts Institute of Technology, and Johns Hopkins University. We also describe the contributions of two notable mathematical pioneers, Benjamin Peirce and Eliakim Hastings Moore. Further information on mathematics in the early colleges can be found in the American Mathematical Society's three-volume *A Century of Mathematics in America*, Karen Parshall and David Rowe's *The Emergence of the American Research Community 1876–1900*, and Florian Cajori's highly detailed *The Teaching and History of Mathematics in the United States,* which covers the story up to 1890.[1]

SOME EARLY COLLEGES

The first institution of higher education to be established in the American colonies was Harvard College in Cambridge, Massachusetts, created in 1636 by the Great and General Court of the Massachusetts Bay Colony. It was named after a Puritan minister, John Harvard of Charleston, Virginia (later, West Virginia), who bequeathed to it his library and half of his estate upon his death in 1638. John Harvard was born in 1607 in London, England, and received his MA degree from Cambridge University in 1635.

Over the following century or so, other colleges and universities were founded. These institutions were privately funded, and their primary aim was to prepare their students for careers in theology, law, medicine, and teaching. After Harvard came

1693: The College of William and Mary in Williamsburg, Virginia

1701: The Collegiate School (later, Yale University) near New Haven, Connecticut

1740: The College of Pennsylvania (later, the University of Pennsylvania) in Philadelphia

1746: The College of New Jersey (later, Princeton University)

1754: King's College (later, Columbia University) in New York City

1764: The College in the English Colony of Rhode Island and Providence Plantations (later, Brown University) in Providence

1769: Dartmouth College in Hanover, New Hampshire

All of these were located in the original thirteen states of the Union, and more were to be founded as time and social development continued.

These early colleges did not include any knowledge of mathematics among their entrance requirements. Later, when mathematics was considered a prerequisite—at Yale in 1745, Princeton in 1760, and Harvard in 1807—it was limited to elementary arithmetic. In 1816, Harvard expected its applicants to have a greater understanding of arithmetic, and it added algebra in 1820. It was not until after the Civil War of 1861–65 that the other colleges insisted on algebra as a prerequisite.

There was indeed little enthusiasm for mathematics in the early years of the American colleges, as illustrated by the low level of the entrance criteria. Members of staff who did not usually teach the subject carried out the instruction of any mathematics that was considered necessary. In 1711, the Reverend Tanaquil Lefevre, the son of a French diplomat, was the first person in the colonies to be appointed a professor of mathematics, at the College of William and Mary. Isaac Greenwood took up a similar position at Harvard in 1726, but neither held his tenure for long. By 1729, the colonies could boast only six professors of mathematics, usually coupled with another subject such as natural philosophy (physics) or astronomy; all of them were graduates of British universities: Oxford, Cambridge, or Edinburgh.

Like all developing nations, America set great store by the education of its population, but the rate of progress, as always, was governed by economics and the caliber of the people who were available to do the teaching. Progress was made over the next century in all areas of learning, not least through the increase in the number of institutions of higher education, with an accompanying considerable growth in the importance of mathematics.

The ending of the Civil War in 1865 initiated a significant increase in the amount of disposable money available, both to the government and to the general populace. This newfound affluence, coupled with the Morrill Act which had been authorized by President Abraham Lincoln in 1862, allowed for a significant increase in the number of American institutions of higher learning. The act was conceived as a vehicle for promoting and enhancing the practical education of the growing industrial population and provided a fundamental change in the perception and funding of higher learning. It allowed also for the income from public land in each state and territory to be allocated to the building of colleges, together with the resources necessary for the teaching of agriculture and mechanical subjects. By making higher education available to more sections of the population than it had been previously—with the express intention that this should be for those who could benefit from further education irrespective of their financial background—the act also increased the admission of women into institutions of higher learning and provided colleges specifically for them and for former slaves and their descendants.

In the history of graph theory, the most significant of these early American colleges were Harvard and Princeton. Later institutions were the Massachusetts Institute of Technology (founded in 1861 in Boston, and later relocated to Cambridge), Johns Hopkins University (founded in 1876 in Baltimore), the University of Chicago (founded in 1890), and the Institute for Advanced Study (established in 1930 and located at Princeton).

Harvard University

To this day, Harvard remains one of the most prestigious institutions of learning in America. It can claim as graduates eight presidents and more than forty Nobel Prize laureates.

When it opened in 1636 with just nine students and a single master, the education provided was based on the European pattern. The college was modeled on traditional English universities with their classic academic

A Prospect of the Colledges in Cambridge in New England, an engraving from 1726.

courses, but was tailored to the Puritan philosophy of the first colonists. Although Harvard was never officially associated with any religious denomination, many of its early graduates entered the Puritan ministry, taking up positions as clergymen throughout New England.

The first non-clergyman to become president of Harvard College was John Leverett in 1708, and under his guidance and that of his successors the curriculum was greatly widened, particularly in the sciences. Indeed, this development was so successful that in 1780 the Massachusetts Constitution officially recognized Harvard as a university. It continued to expand and develop during the early 1800s and acquired a growing reputation.

A Hollis Professorship of Mathematicks and Natural Philosophy was established at Harvard in 1727. It was filled consistently, and mathematics was taught under its aegis throughout the 18th century. Later, under the direction of John Farrar, who held this position from 1807 to 1836, further courses in mathematics became available.

As aids to improving the level of undergraduate instruction, Farrar made translations of 18th-century French works on mathematics, physics, and astronomy. His resignation due to poor health allowed Benjamin Peirce (pronounced "purse"), unquestionably one of the two foremost American mathematicians of the 19th century, to take the leading role in teaching these subjects at Harvard. During the early 1830s he produced new sets of course notes, resulting in further improvements in the quality of teaching.

The Lawrence Scientific School, which opened at Harvard in 1847, was an early attempt to foster graduate education in individual sciences, in place of those previously labeled "natural philosophy" or "natural history". This allowed Peirce the opportunity to develop a graduate program, and from 1836 to the end of the century, mathematics at the university was dominated by Peirce and his students. Indeed, from 1845 to 1865, Harvard became a center of mathematical activity in America, with most of the credit going to Peirce. However, it was not until 1912 that it invested in a dedicated postgraduate program of mathematical research.

Yale University

In 1701, a "Collegiate School" opened in the parsonage of Abraham Pierson of Killingworth, Connecticut, but relocated in 1716 to nearby New Haven, where it has remained ever since. Two years later, following a generous benefaction by Elihu Yale, who had acquired a fortune from the British East India Company, it was renamed Yale College. It did not receive its final designation as Yale University until 1887.

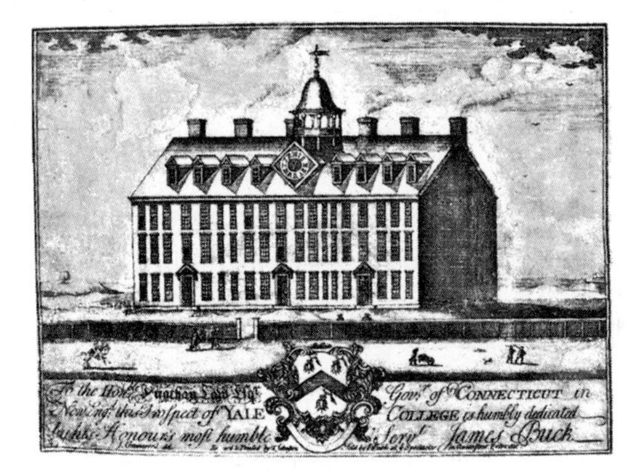

An early view of Yale College.

Among the books presented to the new college in its early days were many mathematical volumes, including Newton's *Principia Mathematica* and *Opticks*. Edmond Halley added his support by sending his edition of Apollonius's *Conics*, and by 1743 the Yale library catalog listed no fewer than fifty-five mathematics books.

The student curriculum at this time was as follows:[2]

> In the first year to study principally the tongues [sacred languages], arithmetic, and algebra; the second, logic, rhetoric, and geometry; the third, mathematics, and natural philosophy; and the fourth, ethics and divinity.

In 1743, Newtonian calculus (fluxions) was added to the third-year offering for those who wished to learn it, and this continued to be taught for many years.

In 1770, a chair of mathematics and natural philosophy was founded. The Reverend Jeremiah Day occupied this position from 1803 to 1820, and in 1817 he was appointed Yale's fifth president while continuing to present his lectures. Because student texts were frequently in short supply, he wrote a number of elementary ones on subjects that ranged from algebra, trigonometry, and geometry to navigation and surveying. These books proved to be highly popular and went into many editions.

The chair of mathematics and natural philosophy was split in 1835, and the mathematics half was occupied by Yale alumnus Anthony D. Stanley until his death in 1853. His replacement was Hubert A. Newton, who had graduated in 1850 and had been a tutor in mathematics since 1852. In a notification to students he announced:[3]

> Students desirous of pursuing the higher branches of mathematics are allowed to choose Analytic Geometry in place of the regular mathematics (Navigation) in the third term of Sophomore year, and Differential and Integral Calculus during the first two terms of Junior year, in place of Greek or Latin.

Newton was promoted to professor at the early age of 25, a position that he held until his death in 1896. After spending an initial year in Paris studying geometry, he returned to Yale with a new interest in astronomy and quickly gained international recognition for his writings on comets and meteors.

Two outstanding students whom Hubert Newton mentored at Yale were Josiah Willard Gibbs, who would become one of the most distinguished American scientists of his time, and Eliakim Hastings Moore, whom we meet again later. After studying philology and mathematics and graduating in 1858, Gibbs switched to engineering and wrote the first American doctoral thesis in the subject, on the design of gears. He became a tutor in mathematics from 1863 to 1866, and was promoted to professor of mathematical physics in 1871. He wrote extensively on a wide range of topics and remained at Yale for the rest of his life.

Princeton University

Princeton was chartered under its original name of the College of New Jersey, by which it was known for its first 150 years. The charter, dated October 22, 1746, was issued by the Province of New Jersey in the name of King George II, and included the pronouncement that "any Person of any religious Denomination whatsoever" may attend. Initially located in Elizabeth, New Jersey, the college moved after a year to Newark, and

Nassau Hall at Princeton.

then to Nassau Hall, Princeton, in 1756. Nassau Hall, one of the largest buildings in the new colonies, was named after King William III, Prince of Orange of the House of Nassau.

The first student body consisted of ten young men who attended classes in the parlor of the Reverend Jonathan Dickinson. In 1780, Princeton's charter was amended so that the trustees were no longer required to swear allegiance to the king of England, and in 1783, the Continental Congress met in Nassau Hall, briefly making Princeton the capital of the newly emerging nation. Nine Princeton alumni attended the Constitutional Convention in 1787, more than from any other American or British institution.

By 1896, the college had developed a sufficiently enhanced educational program that it was granted university status and was renamed Princeton University after its host town. Four years later it opened a graduate school, with mathematics research under the guidance of Henry Burchard Fine. Fine published a number of mathematical research papers on numerical analysis and geometry, but was foremost a writer of textbooks and a gifted leader with skills in administration and the development of academic organizations. In the late 1880s, he had been an active supporter of the founding of the New York Mathematical Society, which became the American Mathematical Society in 1894.

Up to this time, Princeton, in common with most of the American universities, had made little contribution to original mathematics. The graduate school brought some success in research for individual members of the mathematics faculty, although its emphasis remained on teaching. In the early 1900s, Fine was the foremost researcher in mathematics at Princeton, and in order to increase the amount of mathematical research at the university, he took advantage of the preceptorial system at Princeton, newly established by Princeton's president, Woodrow Wilson, to appoint a number of rising scholars as instructors. Two of these were Oswald Veblen (in 1905) and George Birkhoff (in 1909), whom we meet again in Chapter 2. But research continued to be the poor relation with little funding and scant facilities, as later recalled by an early mathematical researcher, Solomon Lefschetz:[4]

> When I came in 1924 there were only seven men there engaged in mathematical research. These were Fine, Eisenhart, Veblen, Wedderburn, Alexander, Einar Hille and myself. In the beginning we had no quarters. Everyone worked at home. Two rooms in Palmer [Laboratory of Physics] had been assigned to us. One was used as a library, and the other for everything else! Only three members of the department had offices. Fine and Eisenhart [as administrators] had offices in Nassau Hall, and Veblen had an office in Palmer.

Although the university's research facilities were less than ideal, much was done between 1924 and 1930 to improve the situation. Fine, in particular, was successful in raising money to support Princeton's growing science programs. After his untimely death in a bicycling accident in 1928, a wealthy trustee and friend funded the construction of a new mathematics building. Completed in 1931, and appropriately named Fine Hall, it also served as the first home of the Institute for Advanced Study with mathematics as its first field of study, and with Albert Einstein among its first members. The improvements in Princeton's facilities also had the effect of enhancing mathematical research in the university.

Massachusetts Institute of Technology (MIT)

On April 10, 1861, a charter was approved to found a school of higher education in Boston. It read "An Act to Incorporate the Massachusetts Institute of Technology, and to Grant Aid to Said Institute and to the Boston Society of Natural History . . .", and four years later it opened to its first students. The efforts to raise funds by the institute's founder and first president, William Barton Rogers, were made more difficult by the outbreak of the Civil War. As a result, classes were initially held in rented accommodations, and it was not until 1866 that the institute's first buildings were completed.

During the early years of this (essentially) engineering school, the head of mathematics was John Daniel Runkle, who had been a pupil and protégé of Benjamin Peirce. Runkle had attended the Lawrence

The Great Dome at MIT.

Scientific School of Harvard, from which he graduated in 1851, and subsequently worked at the Nautical Almanac Office in Cambridge; in both places he enjoyed Peirce's influence and encouragement. Runkle was Rogers's right-hand man, and both were influential in planning and defining the institute's teaching program. Runkle believed that his department was there to provide mathematical instruction for the school's engineering students.

On Runkle's death in 1902, his successor, Harry Walter Tyler, set about making the teaching of mathematics a serious subject in its own right, and not just as a service to engineering. In this enterprise he was encouraged by MIT's president, Richard Maclaurin, who also supported Tyler's efforts to expand the department and to promote mathematical research. The institute relocated to its present site in Cambridge in 1916.

By the 1940s, MIT had established itself in the top division of mathematics research. It earned a reputation for invention, and many successful and significant companies were founded by the institute and by its graduates. Since then, over sixty current or former members have received Nobel Prizes.

Johns Hopkins University

Johns Hopkins University, founded in Baltimore, Maryland, was the first research university in the United States. It was started with funds provided by the American entrepreneur, abolitionist, and philanthropist Johns Hopkins, through a bequest of $7 million, half of which also financed the establishment of the Johns Hopkins Hospital. The university

Gilman Hall at Johns Hopkins University.

opened on February 22, 1876, with Daniel Coit Gilman as its first president. In his installation address, Gilman asked:[5]

> What are we aiming at? . . . The encouragement of research . . . and the advancement of individual scholars, who by their excellence will advance the sciences they pursue, and the society where they dwell.

With the freedom of starting from scratch, and without the need to change entrenched ideas, Gilman set out to create an academic establishment new to America. His guiding premise was to build a research school that, through scholarship, would enhance the general level of human understanding, while improving the individual student's knowledge. He succeeded in developing an atmosphere where teaching and research went hand in hand, and where all faculty members were encouraged to become confident to do both. What Gilman achieved at Johns Hopkins marked a major turning point in American higher education and put forth a challenge to other colleges and universities.

One of Gilman's aims was to make America competitive with Europe, and to help him in achieving this goal he set about recruiting internationally respected scholars with a long history of carrying out research and of encouraging research in others. In particular, to implement this philosophy in the mathematics department, Gilman traveled to Europe in 1875 and secured the services of the British mathematician James Joseph Sylvester, who at 61 years of age was still young enough in mind to be able to instill enthusiasm in young scholars. The story of the development of mathematics at Johns Hopkins University with Sylvester at the helm continues in Chapter 1.

MATHEMATICS EDUCATION

College teaching in the early years of the United States was elementary and followed 18th-century English practice, comprising Latin, Greek, philosophy, and a little mathematics. The last of these included Euclid's *Elements*, the rudiments of trigonometry, some Newtonian mechanics, basic arithmetic, and some algebra.

One consequence of the 1812–15 war between Britain and America over shipping and territory disputes was that most developments in America became more greatly influenced by France than by England, as previously. For mathematicians, this meant looking increasingly toward a country where mathematics and science were held in respect and benefited from considerable support from the government; such an attitude had never been a high priority for the British Parliament. This change in emphasis led to the expansion of mathematics and science faculties within American colleges, and to the creation of additional chairs within these disciplines.

During these early years, little research had been carried out within the higher education system in America. Although some people within this system considered research as prestigious, few facilities were available and there were no internal structures for fostering it. Additionally, because American higher education was almost exclusively devoted to undergraduate teaching, there were little experience and ability available to develop postgraduate study. It was expected that promising graduates should travel to Europe, mostly to Germany, for their doctoral study and research.

The founding of Johns Hopkins University initiated a process of change. Daniel Gilman differed from the presidents of long-established colleges and universities such as Harvard, Princeton, and Yale, which were steeped in entrenched tradition. Recognizing that American higher education lagged far behind that of many European countries—and feeling that, for his new university to survive and grow, it needed to offer an alternative program to that of other institutions—he ensured that Johns Hopkins placed equal emphasis on undergraduate studies and on graduate education that incorporated research and support for technical publication.

The two American mathematicians who made the greatest contributions to the early development of teaching and graduate study in America, and who in their differing ways would influence what was to become graph theory, were Benjamin Peirce of Harvard University, whom we encountered earlier, and Eliakim Hastings Moore of the University of Chicago. We now look at their contributions.

Benjamin Peirce

Throughout the relatively short history of American scholarship, and of mathematics in particular, there have been dynasties, albeit mostly of two-generation duration. One of these was the Peirce family, whose head was Benjamin Peirce. His offspring included his highly acclaimed son Charles Sanders Peirce, mathematician, logician, and philosopher, whom we meet again in Chapter 1.

Benjamin Peirce was born on April 4, 1809, in Salem, Massachusetts. His father was a state legislator in Massachusetts and a librarian at Harvard University. The young Benjamin was educated at Salem Private Grammar School and entered Harvard in 1825, at age 16. After graduating from there in 1829, he taught for two years at George Bancroft's

Benjamin Peirce (1809–80).

Round Hill School in Northampton, Massachusetts, before returning to Harvard as a tutor in mathematics. In 1833, he became professor of mathematics and natural philosophy, and in the same year he received a Harvard master's degree. Although the doctoral degree in today's sense did not exist in the United States in Peirce's time, the first issue of the *American Journal of Mathematics* lists him as a Doctor of Laws (probably an honorary degree). In 1842, he became Perkins Professor of Mathematics and Astronomy at Harvard, a position that he held until his death on October 6, 1880.

Peirce was perhaps the first American-born professor of mathematics to consider research as part of his role, and not just as something to carry out in his spare time. He explored a wide range of research topics, and was instrumental in providing the educational structure that would encourage mathematicians of America to engage in research activity; this would have a considerable influence on those who would develop the subject in the United States. During his time at Harvard, he was influential in elevating the status of the college to that of a leading national institution. As the foremost mathematician and astronomer in the country, he made the first important American contribution to mathematical research, even though the program that he developed at Harvard was so demanding that he averaged only two graduate students per year.

Peirce's research topics included celestial mechanics and the applications of plane and spherical trigonometry to navigation, geodesy, and statistics. Having helped to determine the orbit of Neptune, he was appointed director of longitude determination for the US Coast Survey in 1852, and director of the survey from 1867 to 1874.

He was also interested in algebra and number theory. In 1870, at his own expense, he published *Linear Associative Algebra*, which laid the foundations for a general theory of these algebras and classified all the complex ones with dimension less than 7; in addition, he calculated multiplication tables for more than 150 new algebras. This work was greatly influenced by the work of the Irish mathematician William Rowan Hamilton on quaternions. It was considered to be the first American treatise on modern abstract algebra and the earliest important research to come out of the United States in the area of pure mathematics.

As for his abilities as a teacher and communicator, opinions were divided. During the early part of his career, Peirce wrote and published many student textbooks, on topics that ranged from algebra, plane and solid geometry, and exponential equations and logarithms, to curves, functions, and forces. Although they were well written and contained elegant mathematics, they were generally found to be too demanding for all but the most able. His lecturing style also received criticism, with many of his pupils finding him difficult to follow. As one of them complained:[6]

> I am no mathematician, but that I am so little of one is due to the wretched instruction at Harvard. Professor Peirce was admirable for students with mathematical minds, but had no capacity with others.

But for those pupils who were equipped to appreciate his enthusiasm for mathematics, he proved to be an inspiring teacher:[7]

> We were carried along by the rush of his thought, by the ease and grasp of his intellectual movement. The inspiration came, I think, partly from his treating us as highly competent pupils, capable of following his line of thought even through errors, which reached a result with the least number of steps in the process, attaining thereby an artistic or literary character; and partly from the quality of his mind which tended to regard any mathematical theorem as a particular case of some more comprehensive one, so that we were led onward to constantly enlarging truths.

Eliakim Hastings Moore

The University of Chicago was founded in 1890, largely with money provided by John D. Rockefeller, the noted philanthropist and founder of Standard Oil. From the start, its main objective was the development of postgraduate study and research, together with undergraduate instruction. One mathematician to take advantage of this postgraduate training, as well as to benefit from overseas study, was Eliakim Hastings Moore.

Moore was born on January 26, 1862, in Marietta, Ohio. He attended Woodward High School from 1876 to 1879, and his love of mathematics and astronomy was triggered when he worked for the director of the Cincinnati Observatory during a summer vacation. He enrolled at Yale College at the age of 17, earning his bachelor's degree in 1883 and his doctorate two years later for the thesis *Extensions of Certain Theorems of Clifford and Cayley in the Geometry of n Dimensions.*

Encouraged by his supervisor to continue his studies in Germany, Moore spent the academic year 1885–86 attending the Universities of Göttingen and Berlin. On his return to the United States in 1886, he became a high school instructor for a year and a tutor back at Yale for a further two. In 1889, he took up a position at Northwestern University for three years and then moved to the University of Chicago as professor of mathematics and acting head of the mathematics department. He

Eliakim Hastings Moore
(1862–1932).

became head of the department in 1896, a position that he continued to hold until his death in 1932.

Moore had been recruited by the University of Chicago's first president, the outstanding administrator William Rainey Harper, and together they developed an excellent mathematics faculty. Chicago's doctoral graduates went on to establish and expand many important mathematics departments across America over the first decades of the 20th century. The historian of mathematics, Raymond Clare Archibald, summed him up as follows:[8]

> Moore was an extraordinary genius, vivid, imaginative, sympathetic, foremost leader in freeing American mathematics from dependence on foreign universities, and in building up a vigorous American School, drawing unto itself workers from all parts of the world.

During his forty years at Chicago, Moore devoted considerable time and energy to the building of the mathematical community in America. In 1893, he was instrumental in organizing the International Mathematical Congress at the World's Columbian Exposition in Chicago, held to commemorate the 400th anniversary of the European discovery of the Americas. This attracted the participation of forty-five mathematicians from Austria, Germany, Italy, and nineteen states of the Union, as well as papers from French, Russian, and Swiss scholars, and was the first international mathematics meeting to be held in the United States.

The success of the Chicago congress motivated the New York Mathematical Society, founded in 1888, to become a national society. At Moore's request, it funded publication of the conference proceedings, and on

July 1, 1894, and with his encouragement, it changed its name to the American Mathematical Society. Slowly, other sections of the society began to appear throughout the country; Moore lobbied for one in Chicago, and this section first met in 1897. In 1899, the society introduced a new journal, the *Transactions of the American Mathematical Society*, whose purpose was partly to promote American authors, and elected Moore as its editor in chief. From 1898 to 1900, he served as vice president of the American Mathematical Society, was elected its president in 1901, and in 1906 presented AMS Colloquium Lectures on the theory of bilinear functional operations.

Moore's main areas of research were in algebra, groups, and the foundations of geometry, and in later years he worked on the foundations of analysis. Thirty-one research students earned their doctoral degrees under his supervision, and since then he has had many thousands of doctoral descendants. He received honorary degrees from Göttingen, Yale, Clark, Toronto, Kansas, and Northwestern universities.

Although Moore never worked in graph theory, two of his most successful postgraduate students were to make significant contributions to the subject. These were Oswald Veblen and George Birkhoff, who went on to become leading American mathematicians in the first half of the 20th century, and who form the subject of Chapter 2.

* * * * *

As with many things in America in the 19th and 20th centuries, the development of higher education moved quickly, not least in mathematics. By 1910, and up to the outbreak of World War II, Harvard, Princeton, and Chicago were the leading mathematical establishments in the United States, and were comparable to many European universities. This was due in large part to the lead that had been set by Johns Hopkins University and to the significant changes in graduate education at Harvard in the early 1900s, soon to be followed by other universities such as Princeton and Yale, as well as newly formed institutions such as Clark University in Massachusetts and the University of Chicago. These developments changed the principal emphasis from mathematical education to research, with the work of Benjamin Peirce, and especially the seminal contributions of E. H. Moore in developing postgraduate study, proving to be of immense importance for the future of American mathematics.

In Chapter 1, we develop the story of mathematical research under the influence of J. J. Sylvester and Johns Hopkins University. We also examine the early interest in graph theory in America and meet some of its 19th-century pioneers.

Chapter 1
The 1800s

As we have seen, Johns Hopkins University was the first American educational establishment to be founded with an aim of encouraging and providing facilities for research, and in the fall of 1875 its first president, Daniel Coit Gilman, traveled to Europe to headhunt the very best researchers to lead its departments. Mathematics was the first faculty to open, with James Joseph Sylvester appointed as its guiding light. Sylvester published many papers, including some that relate to graph theory.[1]

The story of Johns Hopkins and its mathematics during its first few years is essentially that of Sylvester, but also involves other notable figures. Two scholars important to its early history, and to the development of mathematics in America, were William Edward Story, a mathematician with a talent for organization but little luck, and Charles Sanders Peirce, a brilliant but somewhat wayward polymath. Also important to our story is Alfred Kempe, a compatriot of Sylvester's, whose erroneous solution of the four color problem was to have a profound influence on graph theory in America over the ensuing years.

JAMES JOSEPH SYLVESTER

J. J. Sylvester was born James Joseph on September 3, 1814, in London. His father, a Jewish merchant, was named Abraham Joseph. In his teenage years, James Joseph added the surname Sylvester, as three names were a necessary requirement for possible migration to America, a step being taken by his brother at the time.

At the age of 14, Sylvester entered University College, London, where he was taught by Augustus De Morgan, the professor of mathematics, but after five months his family decided to withdraw him and send him to study at the Royal Institute School in Liverpool. In 1831, he went to St John's College in the University of Cambridge, but suffered from an illness that caused him to miss most of the academic years 1833–35.

Although a brilliant scholar, coming second in the 1837 Mathematical Tripos examinations, he was not permitted to receive his Cambridge degree because he was Jewish and unwilling to sign the Articles of the Church of England. Indeed, because of his religion, he was unable to gain a university position at either Cambridge or Oxford, even though his undoubted ability deserved such an appointment. However, from 1837 to 1841 he was professor of natural philosophy at University College, London, one of the few non-sectarian institutions, and in 1841 he was awarded bachelor's and master's degrees from Trinity College, Dublin.

In the same year, Sylvester was appointed professor of mathematics at the University of Virginia, but he resigned after only a few months following an unfortunate clash with a student and a lack of support from the university. Unable to obtain another post in America, he reluctantly returned to England where he gained employment as an actuary at a life insurance company in London; he also gave private lessons in mathematics. In 1846, he decided to study law, and during his training as a barrister he met the mathematician Arthur Cayley, whose four-year fellowship at Trinity College, Cambridge, had just ended. Unwilling to take Holy Orders, then a condition of appointment at Trinity, Cayley needed a profession and chose law, studying at Lincoln's Inn in London. Despite their very different personalities, Cayley and Sylvester became lifelong friends and collaborated on many mathematical problems.

In 1855, Sylvester became professor of mathematics at the Royal Military Academy at Woolwich, where he remained until 1870 when War Office regulations required him to retire at age 55. So Sylvester was already retired when in 1876 he received President Gilman's invitation to become the first professor of mathematics at Johns Hopkins University. In September of the previous year, Benjamin Peirce, a friend of Sylvester's, had already written to Gilman to urge him to engage Sylvester:[2]

> Hearing that you are in England, I take the liberty to write you concerning an appointment in your new university, which I think would be greatly for the benefit of our country and of American science if you could make it. It is that of one of the two greatest geometers of England, J. J. Sylvester. If you inquire about him, you will hear his genius universally recognized but his power of teaching will probably be said to be quite deficient. Now there is no man living who is more luminary in his language, to those who have the capacity to comprehend him than Sylvester, provided the hearer is in a lucid

James Joseph Sylvester (1814–97).

> interval. But as the barn yard fowl cannot understand the flight of the eagle, so it is the eaglet only who will be nourished by his instruction . . .
>
> Among your pupils, sooner or later, there must be one, who has a genius for geometry. He will be Sylvester's special pupil—the one pupil who will derive from his master, knowledge and enthusiasm—and that one pupil will give more reputation to your institution than ten thousand, who will complain of the obscurity of Sylvester, and for whom you will provide another class of teachers . . .
>
> I hope that you will find it in your heart to do for Sylvester—what his own country has failed to do—place him where he belongs—and the time will come, when all the world will applaud the wisdom of your selection.

Even though many considered Sylvester to be the finest mathematician in the English-speaking world, he was both surprised and delighted to receive Gilman's invitation to occupy a position from which he would derive considerable enjoyment and success.

On taking up his appointment in May of 1876, at a salary of $5000 per annum (which, at his insistence, was paid in gold),[3] Sylvester set about realizing Gilman's objective by initiating research work in the mathematics department. He selected two graduate fellows, George Bruce Halsted and Thomas Craig, to join the mathematics faculty, and in the fall he recruited William E. Story from Harvard.

Sylvester presented his inaugural lecture on February 22, 1877, the first anniversary of the official opening of the university. His presentation covered many subjects, including how mathematics should be taught and studied, and the role that Johns Hopkins should play in the development of mathematics and of further education in America. He also took the opportunity to attack those English universities that discriminated against all who were not Protestant Christians. Having encountered such prejudice himself, he criticized the damage that had been done to higher education by the exclusion of Jews, Catholics, and others.

In 1861, Yale College became the first American institution to confer doctoral degrees. The first mathematics doctorate there was in 1862, and later degrees were awarded for dissertations on "The Daily Motion of a Brick Tower Caused by Solar Heat" and "On Three-Bar Motion". In the 1870s, doctorates in mathematics were awarded four times at Yale, once at Cornell and at Dartmouth, and twice at Harvard and at Johns Hopkins. Sylvester had been quick to take on postgraduate students, and while in Baltimore he supervised eight of them:[4]

1878: Thomas Craig, *The Representation of One Surface upon Another, and Some Points in the Theory of the Curvature of Surfaces*

1879: George Bruce Halsted, *Basis for a Dual Logic*

1880: Fabian Franklin, *Bipunctual Coordinates*

1880: Washington Irving Stringham, *Regular Figures in n-Dimensional Space*

1882: Oscar Howard Mitchell, *Some Theorems in Numbers*

1883: William Pitt Durfee, *Symmetric Functions*

1883: George Stetson Ely, *Bernoulli's Numbers*

1884: Ellery William Davis, *Parametric Representations of Curves*

Another of Sylvester's preoccupations was the *American Journal of Mathematics*, the oldest mathematics journal in continuous publication in North America, and still being published today. Sylvester is usually credited as its founder, and with the help of William Story he published the first issue in 1878. The *Journal* was intended to be a vehicle for dia-

AMERICAN
Journal of Mathematics
PURE AND APPLIED.

EDITOR IN CHIEF,
J. J. SYLVESTER, LL. D., F. R. S., *Corr. Mem. Inst. of France.*

ASSOCIATE EDITOR IN CHARGE,
WILLIAM E. STORY, PH. D., (*Leipsic.*)

WITH THE CO-OPERATION OF
BENJAMIN PEIRCE, LL. D., F. R. S., PROFESSOR OF MATHEMATICS IN HARVARD UNIVERSITY, In Mechanics,
SIMON NEWCOMB, LL. D., F. R. S., CORR. MEM. INST. OF FRANCE, SUPERINTENDENT OF THE AMERICAN EPHEMERIS, In Astronomy,
AND
H. A. ROWLAND, C. E., In Physics.

PUBLISHED UNDER THE AUSPICES OF THE
JOHNS HOPKINS UNIVERSITY.
Πάντα γα μὰν τὰ γιγνωσκόμενα ἀριθμὸν ἔχοντι.—Philolaos.

Volume I.

BALTIMORE:
PRINTED FOR THE EDITORS BY JOHN MURPHY & CO.

B. WESTERMANN & CO., D. VAN NOSTRAND, *New York.*
TRÜBNER & CO., *London.*
A. WILLIAMS & CO., *Boston.*
FERREE & CO., *Philadelphia.*
A. ASHER & CO., *Berlin.*
GAUTHIER-VILLARS, *Paris.*
1878.

The first issue of the *American Journal of Mathematics*, 1878.

log between American mathematicians, although space was also made available for foreign contributions. Indeed, the first issue included contributions from the Americans Simon Newcomb, C. S. Peirce, William Story, Thomas Craig, George Halsted, and Fabian Franklin, while other contributing authors were the Englishmen Arthur Cayley, William Kingdon Clifford, Edward Frankland, and Sylvester himself.

The first six volumes of the *Journal*, which covered 1878–83 and for which Sylvester was responsible, contained nearly two hundred articles. Papers by Sylvester featured in each volume, with thirty-two entries in total, and Cayley contributed to five of these volumes. Another early European contributor was the Danish mathematician Julius Petersen (see Interlude A), while other Americans included Benjamin Peirce and the rest of Sylvester's doctoral students. Moreover, Sylvester had been successful in promoting the new publication, with its "List of Subscribers" on July 1, 1878, totaling nearly 150; thirty-six of these were institutions, some of which took multiple copies, with three addresses in Paris, six in England, and two in Canada.

Sylvester was happy at Johns Hopkins University. For the first time in his life, he was able to teach and carry out research based on his own ideas and on chosen topics within a university environment. His *Mathematical Seminarium*, as he called his school of mathematics, was soon recognized in American mathematical circles and in Europe, while papers published by this group, most of which appeared in the *American Journal of Mathematics*, were widely read at home and abroad. The American mathematician George Andrews has commented that the collective output during these years amounted to a "monumental" contribution to combinatorics,[5] and it was widely accepted that Sylvester and his school were succeeding in putting America on the mathematical map.

In December of 1879, the university issued the first of its *Johns Hopkins University Circulars*. This publication was initially intended to communicate the full scope of the research being undertaken throughout the university; indeed, Sylvester published some of his notes, papers, and lectures there. It also included correspondence between (and information about) members of the various faculties, and in a letter to Cayley in 1883, Sylvester observed that the *Circulars* acted as "a sort of record of progress in connection with the work and personality of the Johns Hopkins".[6]

Chemistry and Algebra

William Kingdon Clifford, a graduate of Trinity College, Cambridge, was one of the major British mathematicians of his time before his untimely death at the age of 33. Clifford believed, as did Sylvester, that there were direct connections between chemistry and the algebra of invariants.

Edward Frankland was a British chemist who held appointments in Britain and in continental Europe, and who for many years was responsible for the continuous analysis of London water supplies; he also served on a Royal Commission on water pollution. In 1866, based on the chemical theory of valency that had recently been introduced by August Kekulé and others, Frankland published his introductory *Lecture Notes for Chemical Students*,[7] in which he explained how atoms and bonds could be depicted graphically with circles and connecting lines. Beginning with water, he also listed the "symbolic formulae" and "graphic notations" for several chemical compounds. His symbolic formulae were expres-

Edward Frankland (1825–99) and William Kingdon Clifford (1845–79).

sions of the atoms and their quantities which combine to form chemical compounds, and for his graphic notation he represented each atom by a letter enclosed in a circle, with all single and multiple bonds identified by lines joining the appropriate circles. For example, he gave water the symbolic formula $\mathbf{OH_2}$ to indicate an oxygen atom (with valency 2) linked to two hydrogen atoms, and his symbolic formula for "ammonic chloride" was $\mathbf{NH_4Cl}$, with a nitrogen atom (with valency 5) linked to a chlorine atom and four hydrogen atoms.

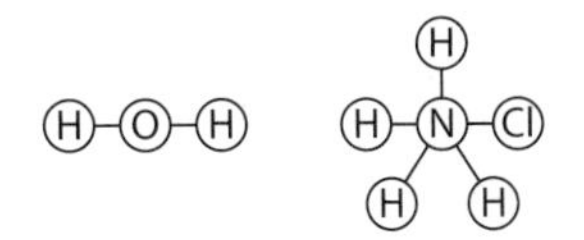

Frankland's graphic notations for water (H_2O) and ammonium chloride (NH_4Cl).

Sylvester was already convinced of the connection between chemistry and algebra and was much taken with Frankland's *Lecture Notes*. In 1878, while at Johns Hopkins, Sylvester wrote a short note that was published in *Nature*.[8] Its opening paragraph shows his enthusiasm for the subject, and the extent to which he had been energized by Frankland:

> It may not be wholly without interest to some of the readers of *Nature* to be made acquainted with an analogy that has recently forcibly impressed

me between branches of human knowledge apparently so dissimilar as modern chemistry and modern algebra. I have found it of great utility in explaining to non-mathematicians the nature of the investigations which algebraists are at present busily at work upon to make out the so-called *Grundformen* or irreducible forms appurtenant to binary quantics taken singly or in systems, and I have also found that it may be used as an instrument of investigation in purely algebraical inquiries. So much is this the case that I hardly ever take up Dr. Frankland's exceedingly valuable *Notes for Chemical Students*, which are drawn up exclusively on the basis of Kekulé's exquisite conception of *valence*, without deriving suggestions for new researches in the theory of algebraical forms. I will confine myself to a statement of the grounds of the analogy, referring those who may feel an interest in the subject and are desirous for further information about it to a memoir which I have written upon it for the new *American Journal of Pure and Applied Mathematics*, the first number of which will appear early in February.

This note was typical of Sylvester's writing style—scholarly, but verging on the flowery. As promised, he then expanded on this note in a paper in Volume I of his *American Journal of Mathematics*.[9]

J. J. Sylvester: *On an application of the new atomic theory to the graphical representation of the invariants and covariants of binary quantics,—with three appendices* (1878)

In this lengthy paper, Sylvester described in detail his reasons for believing in a close connection between the chemistry of organic molecules and the algebraic study of invariant theory. Its first two paragraphs give a flavor of his prose, in language that one now rarely encounters in academic papers:

By the *new* Atomic Theory I mean that sublime invention of Kekulé which stands to the *old* in a somewhat similar relation as the Astronomy of Kepler to Ptolemy's, or the System of Nature of Darwin to that of Linnaeus;—like the latter it lies outside of the immediate sphere of energetic, basing its laws on pure relations of form, and like the former as perfected by Newton, these laws admit of exact arithmetic definitions.

Casting about, as I lay awake in bed one night, to discover some means of conveying an intelligible conception of the objects of modern algebra to a mixed society, mainly composed of physicists, chemists and biologists, interspersed only with a few mathematicians, to which I stood engaged to give some account of my recent researches in this subject of my predilection, and impressed as I had long been with a feeling of affinity if not identity of object between the inquiry into compound radicals and the search for "Grundformen" or irreducible invariants, I was agreeably surprised to find, of a sudden, distinctly pictured on my mental retina a chemico-graphical image serving to embody and illustrate the relations of these derived algebraical forms to their primitives and to each other which would perfectly accomplish the object I had in view, as I will now proceed to explain.

In this paper he again heaped praise on Frankland's *Lecture Notes*:

The more I study Dr Frankland's wonderfully beautiful little treatise the more deeply I become impressed with the harmony or homology (I might call it, rather than analogy) which exists between the chemical and algebraical theories.

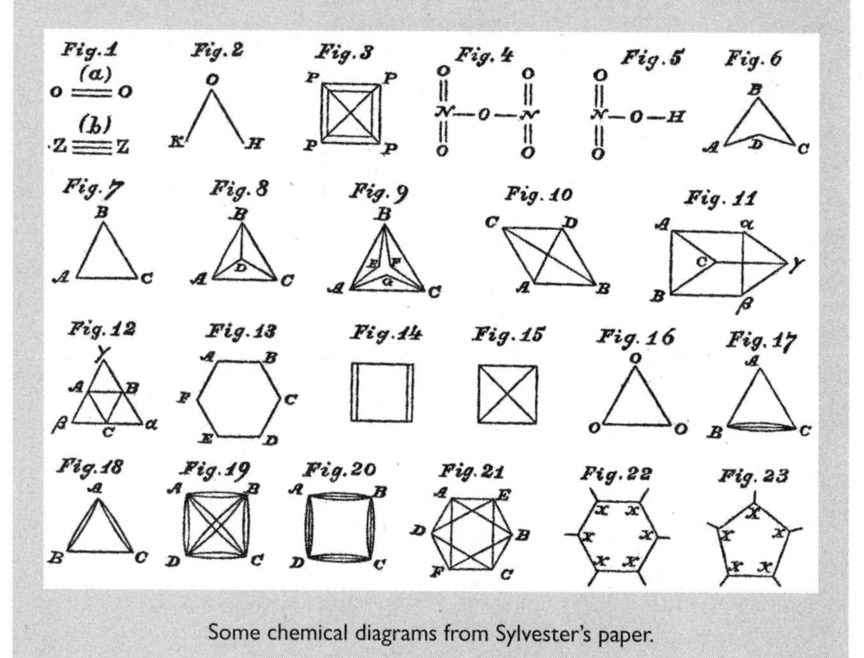

Some chemical diagrams from Sylvester's paper.

Later in the same work he became even more eloquent, enthusing that "I feel as Aladdin might have done in walking in the garden where every tree was laden with precious stones", and continuing:

> Chemistry has the same quickening and suggestive influence upon the algebraist as a visit to the Royal Academy, or the old masters may be supposed to have on a Browning or a Tennyson. Indeed it seems to me that an exact homology exists between painting and poetry on the one hand and modern chemistry and modern algebra on the other. In poetry and algebra we have the pure idea elaborated and expressed through the vehicle of language, in painting and chemistry the idea enveloped in matter, depending in part on manual processes and the resources of art for its due manifestation.

The analogy that Sylvester was trying to make was between "binary quantics" in algebra and atoms in chemistry. A *binary quantic* is a homogeneous expression in two variables, such as

$$ax^3 + 3bx^2y + 3cxy^2 + dy^3,$$

and an *invariant* is a function of the coefficients a, b, c, and d that remains essentially unaltered under linear transformations of the variables x and y. Sylvester explained that this analogy evolved from his diagrammatic representations of chemical compounds, and in his 1878 note in *Nature*, he provided the following explanation of the connections between atoms and binary quantics. It is here that the word *graph* (in our modern sense) made its first appearance.

> The analogy is between atoms and *binary* quantics exclusively.
>
> I compare every binary quantic with a chemical atom. The number of factors (or rays, as they may be regarded by an obvious geometrical interpretation) in a binary quantic is the analogue of the number of *bonds*, or the *valence*, as it is termed, of a chemical atom.
>
> Thus a linear form may be regarded as a monad atom, a quadratic form as a duad, a cubic form as a triad, and so on.
>
> An invariant of a system of binary quantics of various degrees is the analogue of a chemical substance composed of atoms of corresponding *valences*.

> The order of such [an] invariant in each set of coefficients is the same as the number of atoms of the corresponding *valence* in the chemical compound . . .
>
> The weight of an invariant is identical with the number of the bonds in the chemicograph of the analogous chemical substance, and the weight of the leading term (or basic differentiant) of a co-variant is the same as the number of bonds in the chemicograph of the analogous compound radical. Every invariant and covariant thus becomes expressible by a *graph* precisely identical with a Kekuléan diagram or chemicograph . . . I give a rule for the geometrical multiplication of graphs, that is, for constructing a *graph* to the product of in- or co-variants whose separate graphs are given.

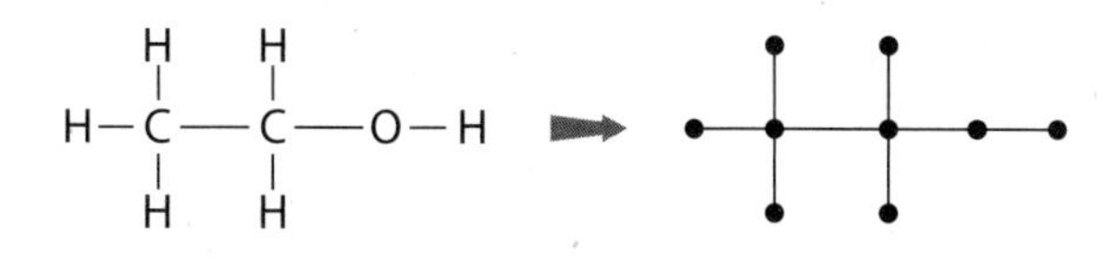

The graph of a chemical molecule.

In spite of his enthusiasm for his analogy between chemistry and algebra, Sylvester was somewhat apprehensive that it might not meet with universal acceptance. Perhaps he suspected that it would be rejected, as he wrote to Simon Newcomb, a mathematician and astronomer at the Naval Observatory in Washington, who in 1884 became professor of mathematics and astronomy at Johns Hopkins:[10]

> I feel anxious as to how it will be received as it will be thought by many strained and over-fanciful. It is more a "reverie" than a regular mathematical paper. I have however added some supplementary mathematical matter which will I hope serve to rescue the chemical portion from absolute contempt. It may at the worst serve to suggest to chemists and Algebraists that they may have something to learn from each other.

Although there was some academic debate on the theory, it soon ran its course as it became apparent that the only link between chemistry and algebra was "the use of a similar notation".[11] Despite the detailed descriptions in Sylvester's note, the associated paper, and his subsequent correspondence with chemists and mathematicians, his ideas were generally considered to have only a passing connection between Kekulé's notation for chemical compositions and the theory of trees developed by Arthur Cayley.

Trees

A *tree* is a connected graph without cycles. In any tree the number of edges is one less than the number of vertices, and any connected graph with this property is a tree.

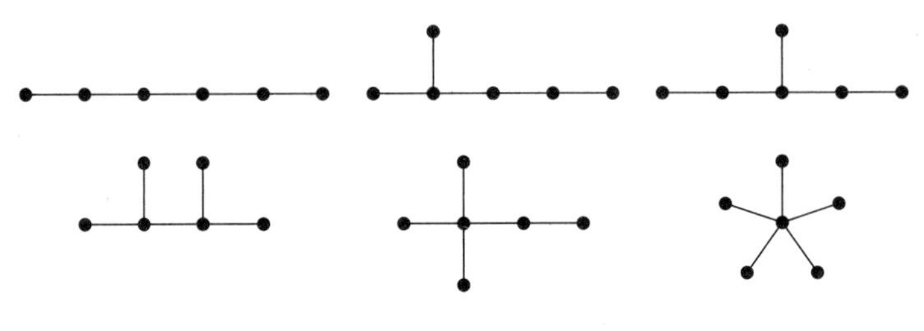

The trees with six vertices.

As we have seen, Cayley had met Sylvester during their years in London, and they remained lifelong friends and collaborators on mathematical matters. Between 1857 and 1889, Cayley produced a number of publications on trees. His first paper of 1857 was the earliest to use the word "tree" in our sense,[12] although both Gustav Kirchhoff (in connection with his work on electrical networks) and Karl Georg Christian von Staudt had used the idea around ten years earlier. Cayley's interest in trees originally "arose . . . from the study of operators in the differential calculus", being inspired by some of Sylvester's work on "differential transformation and the reversion of serieses". His earliest papers dealt with *rooted trees* only, in which one particular vertex is designated as the "root", usually placed at the top, as follows:

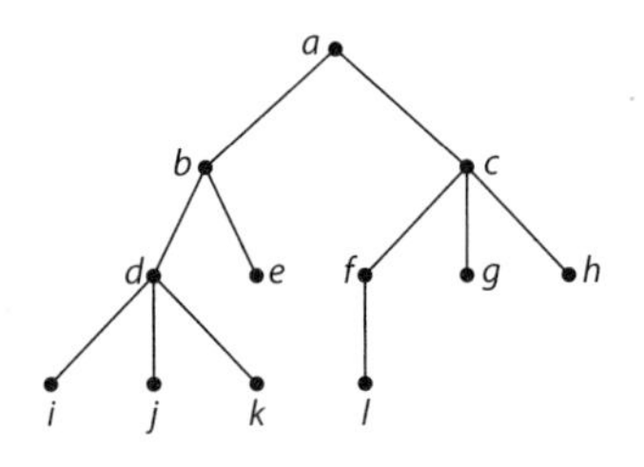

Isomers are chemical compounds with the same chemical formula but different atomic configurations; the next figure shows two molecules with the formula C_4H_{10} (n-butane and 2-methyl propane, formerly called butane and isobutane). Cayley wrote several papers in which he related work on chemical compositions to his studies of trees, and in 1874 he published the short paper "On the mathematical theory of isomers",[13]

in which his work on trees was used in the recognition and enumeration of chemical isomers. Two further papers, in 1875 and 1877, also dealt with the connections between trees and chemical composition.[14]

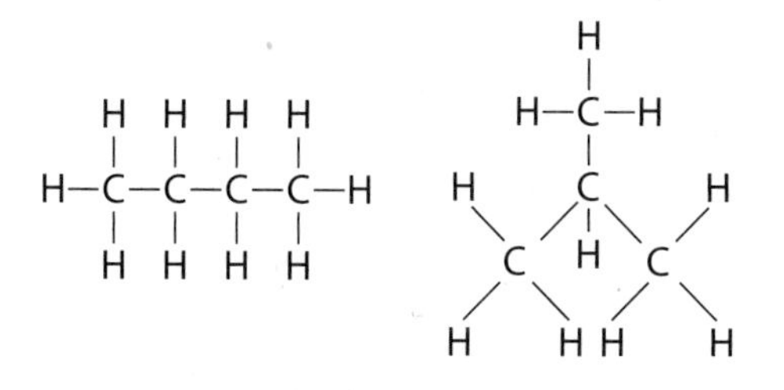

Two chemical isomers.

Sylvester wrote two short papers on trees while at Johns Hopkins. The first of these, "On the mathematical question, what is a tree?", was published in 1879 in the *Mathematical Questions with Their Solutions, from the "Educational Times"*. The second, on "ramifications" (his name for trees), appeared in the same year in the first issue of the *Johns Hopkins University Circulars*.[15]

Sylvester undoubtedly felt the lack of mathematical peers at Johns Hopkins University and in the United States generally, especially after the death of Benjamin Peirce in 1880, and wished that Cayley could visit him. Deprived of their frequent meetings of earlier days, Sylvester sent

Arthur Cayley (1821–95).

him a number of letters in early 1881, inviting him to teach for a period at Johns Hopkins. Sylvester painted an encouraging picture of the social and academic life in Baltimore, and promised that Cayley would be rewarded both academically and financially. His letters, and a visit to Cayley in Cambridge in August 1881, finally persuaded Cayley to visit Johns Hopkins for six months during the spring semester of 1882, and to present a series of lectures during his visit. While there, Cayley also published papers in the *Johns Hopkins University Circulars* and the *American Journal of Mathematics*.

ALFRED KEMPE

In 1852, Francis Guthrie, a former student of Augustus De Morgan's at University College, London, was coloring the counties of a map of England. Finding that just four colors were sufficient for this task, he asked the following more general question, which would become known as the *four color problem*:

> Can the countries of every map be colored with at most four colors so that no two neighboring countries are colored the same?

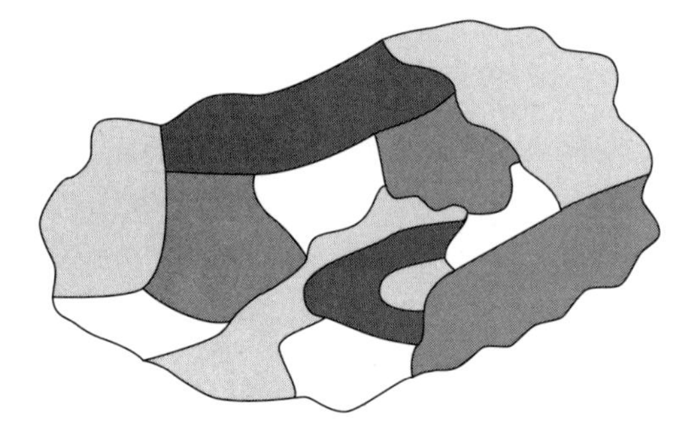

A map that requires four colors.

De Morgan became intrigued by the problem and wrote to the Irish mathematician William Rowan Hamilton and others asking whether four colors always suffice. He also mentioned it in a review of a book by William Whewell in *The Athenaeum*,[16] but died in 1871 without knowing the answer. The *four color theorem*, that all maps on the plane or a sphere can indeed be so colored, was not proved until 1976—by Kenneth Appel and Wolfgang Haken, two mathematicians at the University of Illinois at Urbana–Champaign (see Chapter 6).[17]

Arthur Cayley also became interested in the four color problem, and on June 13, 1878, at a meeting of the London Mathematical Society, he asked whether it had been solved; his query was recorded in the society's *Proceedings* and in a report of the meeting in *Nature*.[18] In a short note in the *Proceedings of the Royal Geographical Society* in April 1879,[19] he described some of the difficulties inherent in tackling the problem. He also made the useful suggestion that certain restrictions can be imposed on the maps under consideration without any loss of generality; in particular, he proved that when tackling the four color problem, we may assume that they are *cubic maps*, with exactly three countries at each meeting point. From now on, when desirable, we shall assume that the maps we are considering are cubic maps.

Alfred Bray Kempe (1849–1922).

Also attending the London Mathematical Society's meeting was Alfred Kempe (pronounced "kemp"), a former student of Cayley's at Cambridge, and yet another English mathematician who then became a barrister. Most of Kempe's early mathematical work was associated with the geometry of mechanical linkages. He later became treasurer of the Royal Society of London and was knighted in 1912.

Kempe was intrigued by Cayley's query on the four color problem and believed that he could solve it. On July 17, 1879, he announced a "solution" in *Nature*.[20] His attempted proof of the four color theorem was "On the geographical problem of the four colours" and—presumably at Cayley's suggestion—he submitted it to the newly founded *American Journal of Mathematics,* which was seeking papers from European authors. Kempe outlined the inherent challenge as follows:[21]

> Some inkling of the nature of the difficulty of the question, unless its weak point be discovered and attacked, may be derived from the fact that a very small alteration in one part of a map may render it necessary to recolour it throughout. After a somewhat arduous search, I have succeeded, suddenly, as might be expected, in hitting upon the weak point, which proved an easy one to attack. The result is, that the experience of the map-makers has not deceived them, the maps they had to deal with, viz: those drawn on simply connected surfaces, can, in every case, be painted with four colours. How this can be done I will endeavour—at the request of the Editor-in-Chief—to explain.

As we have seen, the editor in chief was J. J. Sylvester.

Kempe's paper was published later in the year in Volume 2 of the *American Journal of Mathematics*. Unfortunately, it contained a fatal error which was not discovered until eleven years later, during which time his proof had become generally accepted. In 1890 Percy Heawood exposed Kempe's error (see Interlude A).

A. B. Kempe: *On the geographical problem of the four colours* (1879)

In 1750, Leonhard Euler observed that if a polyhedron has F faces, E edges, and V vertices, then $F+V=E+2$. Using this result, Kempe deduced that if a map has D districts or countries (not counting the external region), B boundaries between countries, and P "points of concourse" where at least three districts meet, then

$$P+D-B-1=0.$$

He then used a counting argument to show that, for a general map,

$$5d_1+4d_2+3d_3+2d_4+d_5-\text{etc.}=0,$$

where, for each k, d_k is the number of districts of the map with k boundaries, and the term "etc." is a collection of terms whose sum is positive. It follows that the sum of the first five terms is also positive, and so not all of d_1 to d_5 can be 0—that is:

> Every map drawn on a simply connected surface must have a district with less than six boundaries.

(A surface is *simply connected* if it is in one piece and has no "holes", so a plane or sphere is simply connected but a torus is not.)

From this remarkable result, Kempe developed an algorithm for coloring any map by using a system of what he called "patches". This process involved selecting a district with five or fewer neighbors and covering it with a slightly larger blank piece of paper, or patch. He then joined all the boundaries that touch the edge of the patch to a single point within the patch; this has the effect of reducing the number of districts by 1, as shown below. The process is then repeated until only one district remains—as Kempe put it, "The whole map is patched out"—and this remaining district is then colored with any of the four colors.

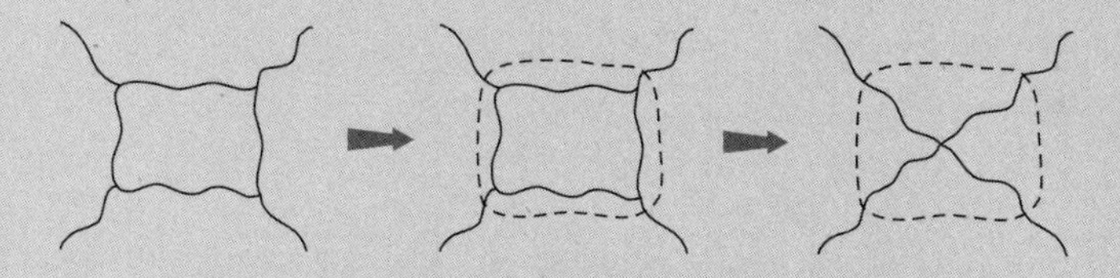

Kempe's patching process.

Kempe then reversed the patching process, removing one patch at a time and successively coloring the uncovered districts with any available color until the original map was colored with four colors. Unfortunately, his explanation of this final step was incomplete. His patching procedure works as long as each restored district has at most three boundaries, but if it has four or five boundaries, then it may be surrounded by districts that require all four colors.

To overcome this difficulty, Kempe developed a strategy for coloring maps that is now called the *method of Kempe chains* or a *Kempe-chain argument*. In this method, we interchange two colors in order to enable the coloring of two neighboring districts that could not previously be colored. His argument was based on the fact that if we are given a map in which all the districts except one are colored, and if the districts that surround the uncolored one are assigned all four colors, then such an interchange of colors can enable the uncolored district to be colored also. This important line of argument was later to become one of the standard tools in the coloring of maps.

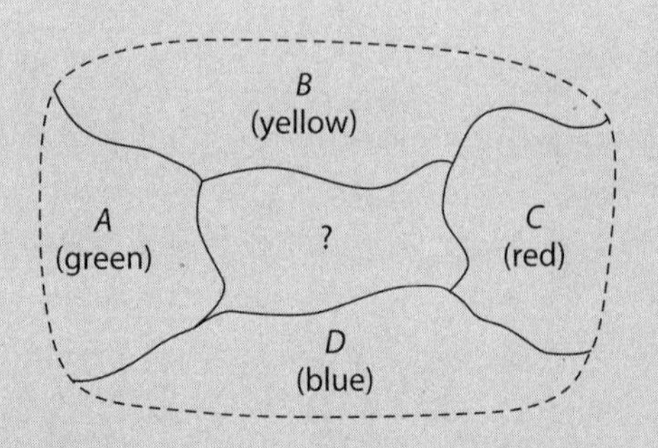

As an example of his method, consider the above map in which the uncolored district is surrounded by four districts that have been assigned different colors. The districts A and C that have been assigned the colors *green* and *red* are either connected by a continuous chain of *green* and *red* districts, or are not so connected. In the latter case, we may exchange the colors *green* and *red* in the chain of *green–red* districts connected to district A without altering the color of district C; this exchange of colors results in districts A and C both being colored *red*, so that the uncolored district can be colored *green*. However, if there is a continuous chain of *green–red* districts that joins A and C, then there is no advantage to making such an interchange of colors. But in this case there can be no continuous chain of *yellow* and *blue* districts joining districts B and D, and so we can recolor either of the *yellow–blue* chains connected to B or D. The districts B and D are then either both *yellow* or both *blue*, thereby allowing the four surrounding districts to be colored with three colors, and leaving the fourth color for the central one.

Kempe then considered maps containing an uncolored district with five sides, and the incorrect application of his method in this case gave rise to his famous error. His mistake was to attempt two color interchanges at the same time; either interchange by itself would have been valid, but to apply them simultaneously could result in two neighboring districts receiving the same new color.

Kempe then noted the following two special cases of interest:

If there is an even number of boundary lines at each point, two colors suffice to color the map.

If every district of a cubic map has an even number of boundary lines, three colors suffice.

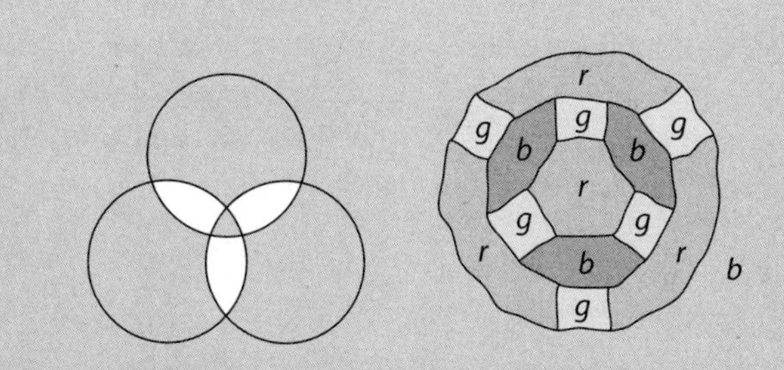

Before leaving his paper, we mention that Kempe was the first to introduce the idea of *duality* and to pose the dual formulation of the four color problem:

> If we lay a sheet of tracing paper over a map and mark a point on it over each district and connect the points corresponding to districts which have a common boundary, we have on the tracing paper a diagram of a "linkage," and we have as the exact analogue of the question we have been considering, that of lettering the points in the linkage with as few letters as possible, so that no two directly connected points shall be lettered with the same letter.

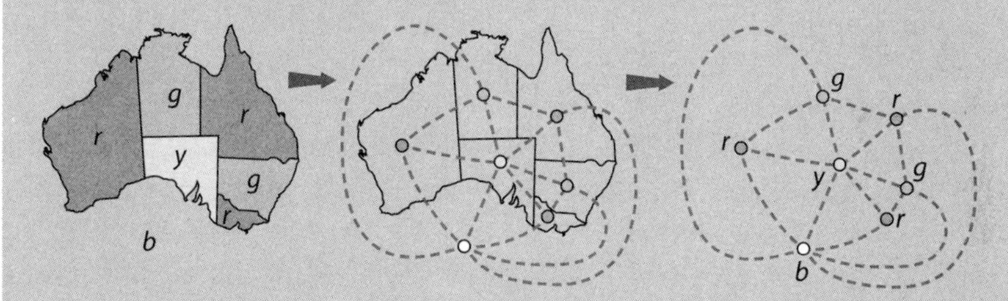

Coloring a map and its dual graph.

So the four color problem can be equivalently stated in terms of coloring the points of a related linkage or graph, as just explained. This idea will reappear in Chapter 4, where we investigate duality, and Chapter 6, where we describe the eventual resolution of the problem.

In the subsequent months, Kempe issued two revisions of his proof.[22] The first, an untitled abstract that is "simpler, and is free from some errors which appeared in the former" was published in the *Proceedings of the London Mathematical Society* in 1879. There, Kempe mentioned his proof in the *American Journal of Mathematics*, and provided a streamlined description of his reduction and patching methods, with instructions for interchanging colors within chains. His second follow-up paper, "How to colour a map with four colours", published in *Nature* on February 26, 1880, was similar in content to the untitled abstract. It was again offered as a simplification, and in Kempe's own words,

> I have succeeded in obtaining the following simple solution in which mathematical formulae are conspicuous by their absence.

Neither of these revised versions indicated any recognition of his fundamental error.

On reading these two papers today, we cannot help arriving at the conclusion that Kempe was not trumpeting his claimed achievement, but was modestly confident that he had found the solution to a problem that had vexed and entertained a considerable number of mathematicians, both professional and amateur.

WILLIAM STORY

William Edward Story was born on April 29, 1850, in Boston, Massachusetts. One ancestor, the Englishman Elisha Story, had arrived in America around 1700 and settled in Boston, while another took part in the Boston Tea Party.

William Story entered Harvard University in 1867, and was one of the first students to be awarded the newly created honors degree in mathematics. He then became one of the earliest American mathematicians to attend a German university, gaining a doctorate from Leipzig University

William Story
(1859–1930).

in 1875 for his dissertation, *On the Algebraic Relations Existing Between the Polars of a Binary Quantic.*

On returning to the United States, Story became a tutor at Harvard University. He is known to have impressed Benjamin Peirce while an undergraduate at Harvard, and this view increased as Story carried out his tutorial duties. Indeed, so convinced was Peirce of Story's merits, that when Sylvester solicited suggestions of suitable mathematicians to join the newly founded department of mathematics at Johns Hopkins University, Peirce recommended Story.

In the hot summer months of 1876, Sylvester decided to return to England, and it was left to the Johns Hopkins president, Daniel Gilman, to interview Story and to make any decision on his employment as Sylvester's assistant. Gilman's initial terse approach was not enthusiastically received by Story, who found it a little patronizing, and his reply was perhaps a trifle sharp. But he did ask for an interview, during which he outlined his ideas for the creation of a learned mathematical journal and a student society. Story was duly offered the Johns Hopkins position, but not before he had tried unsuccessfully to improve his status at Harvard. In the autumn, Story moved to Baltimore as an "associate" (equivalent to an assistant professor at some other universities). Later, in 1883, when the university introduced the title of associate professor, Story was promoted to that position.

Initially, life at Johns Hopkins went well for Story. He set about helping to develop the mathematics department, and his preference was to model it on the example he had experienced while in Germany. He assisted Sylvester in setting up the *American Journal of Mathematics* and was intimately involved in the founding of a mathematical society within the university. As Roger Cooke and V. Frederick Rickey have observed:[23]

> There is evidence that Story succeeded in founding his student mathematical society. *The Johns Hopkins University Circulars*, which are a rich source of information about the university, contain titles and reports of the talks given at the monthly meeting of the "Mathematical Society." From one of these we learn that when Lord Kelvin lectured at Hopkins in 1884, he spoke to a group of mathematicians who called themselves "the coefficients".

Because Sylvester was not good with either finance or management, he appointed Story as associate editor in charge of the *Journal*, and soon praised his second-in-command in a letter to Benjamin Peirce:[24]

> Story is a most careful managing editor and a most valuable man to the University in all respects and an honor to the University and its teachers from whom he received his initiation.

However, the way in which the *Journal* was run was soon to cause friction between Story and Sylvester. This was not a personal difference, but a dissimilarity in the ways that they believed the journal should be edited. During his time at Johns Hopkins, it was Sylvester's custom to spend each summer in England, leaving America in the late spring and returning for the start of the next academic year, while Story was left in charge for the duration of Sylvester's annual leave. The situation came to a head during Sylvester's absence from America through the publication of Kempe's paper on the four color problem.

Story had reviewed Kempe's paper, and on November 5, 1879, he presented the salient points of the "proof" to an audience of eighteen at a meeting of the Johns Hopkins Scientific Association. He then offered "a number of minor improvements", which he put in the form of a note that "was intended to make the proof absolutely rigorous". Story's "Note on the preceding paper" was then published in the *American Journal of Mathematics*, immediately following Kempe's.[25] By presenting it, Story was to incur the wrath of Sylvester, as we shall see.

In his note, Story addressed special cases that Kempe had not covered in his paper. He used both Euler's formula and the patch method, as Kempe had done, but endeavored to be more precise in his use of the formulas contained in Kempe's paper. Story's opening paragraph set out his intention, saying:

> it seems desirable, to make the proof absolutely rigorous, that certain cases which are liable to occur, and whose occurrence will render a change in the formulae, as well as some modification of the method of proof, necessary, should be considered separately.

It is disappointing that Story was not able to identify the major flaw in Kempe's argument in his review of Kempe's paper and in developing his own contribution.

W. E. Story: *Note on the preceding paper* (1879)

Story concentrated on two parts of Kempe's paper. The first of these expanded on the patching method, as applied to three of Kempe's figures, and the second dealt with the cases in which more than three districts meet at a point.

At each stage of the patching, Kempe had denoted the number of districts by D, the number of boundaries by B, and the number of points by P, and had used the corresponding symbols D', B', and

P' after the next patch was removed. Story took up the argument that if the next patch had no point or boundary on it when it was removed, then an island would appear. Following Kempe, he concluded that, in this case,

$$P' = P, \quad D' = D + 1, \quad \text{and} \quad B' = B + 1.$$

However, if the patch had no point but only a single boundary, so that a peninsula or a district with two boundaries appeared when the patch was removed, then for the peninsula,

$$P' = P + 1, \quad D' = D + 1, \quad \text{and} \quad B' = B + 2,$$

and for the district with two boundaries,

$$P' = P + 2, \quad D' = D + 1, \quad \text{and} \quad B' = B + 3.$$

In the second case, Story referred to Kempe's Figure 15.

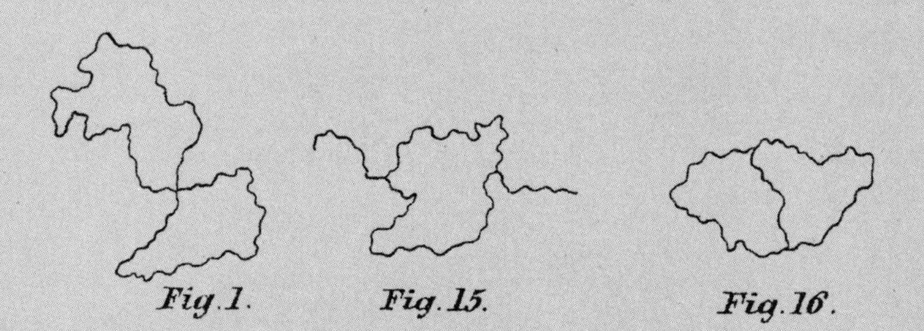

He went on to assert that

> These formulae hold only if the boundaries joined by the line on the patch counted as *two* (and not *one*, as in Figs. 16 and 1) before the patch was put on.

Story then considered a point where boundaries met, and where a district with β boundaries appeared, when the patch was removed. This gave

$$P' = P + \beta - 1, \quad D' = D + 1, \quad \text{and} \quad B' = B + \beta.$$

Story deduced that these equations were identical to those of Kempe (although Kempe had used σ, rather than β),

> only when three and no more boundaries meet in each point of concourse about the district patched out,

giving

$$P' + D' - B' - 1 = P + D - B - 1.$$

Story continued by detailing the alternative situation where the patch has no point of concourse, but only a single line that formed part of the boundary of a district or an island. Removing the patch then revealed Kempe's Figure 16 or Figure 1.

For the district,

$$P' = P + 1, \quad D' = D + 1, \quad \text{and} \quad B' = B + 1,$$

and for the island,

$$P' = P + 2, \quad D' = D + 1, \quad \text{and} \quad B' = B + 2,$$

and so in both cases,

$$P' + D' - B' - 1 = P + D - B.$$

Story next defined a *contour* as an aggregate of boundaries, with the contour being simple or complex, according to whether it contained one, or more than one, district. He asserted that one could improve Kempe's theorem by including contours in the patching procedure. In particular, where Kempe had stated that:

> in every map drawn on a simply connected surface the number of points of concourse and number of districts are together one greater than the number of boundaries,

Story's theorem read:

> in every map drawn on a simply connected surface the number of points of concourse and number of districts are together one greater than the number of boundaries and number of complex-contours together.

As he explained:

> If then x of the contours formed by the boundaries of any map are complex, for that map
>
> $$P+D-B-1=x.$$

In the second half of his paper, Story questioned one of Kempe's claims that

> if we develop a map so patched out, since each patch, when taken off, discloses a district with less than six boundaries, not more than five boundaries meet at the point of concourse on the patch.

He asserted that this is valid only when the number of boundaries meeting in each point does not exceed 3, and detailed a procedure to overcome this restriction. His solution was to use an auxiliary patch whenever more than three boundaries met, thereby reducing to 3 the number of boundaries at a point; one could then continue the method of patching as described by Kempe. On completing the patching and arriving at a map with just one district and no boundary, coloring could then commence as the map was developed by removing patches (including auxiliary patches) in reverse order. By this method, he maintained, "the map will be coloured with four colours", as required.

Sylvester versus Story

Sylvester believed that there had been an undue delay in the publication of Volume 2 of the *American Journal of Mathematics* during his absence in England. He also complained that previously agreed editorial decisions had been changed, and that Story should not have published his note. Sylvester went on to call this "unprofessional", and the relationship between the two colleagues became strained.

In 1880, Sylvester wrote to President Gilman, protesting Story's "conduct" and his "disobeying my directions". In June, he wrote again asking why Story had not sent him an acknowledgment regarding a paper that Sylvester had sent from England. Then, still aggrieved, Sylvester sent a further letter of eight pages to Gilman on July 22—indeed, such was his annoyance that his haste made parts of the letter even more illegible than usual.[26] In this letter, Sylvester complained that he was not told whether the *Journal* had been published and, if so, when. He also objected to his treatment by Story and questioned whether other contributors had been dealt with in an equally poor manner. Sylvester no longer had confidence in Story and was so incensed that he formally requested that Story should have no further involvement with the *Journal*. He also made it clear that Story could be made aware of his opinion and the contents of his letter.

Gilman mediated between the two, but Story's name did not appear on later issues of the *Journal*. Story resigned from the editorial board and began to seek a new position, a task that took him several years to accomplish. As with most disagreements, it would be wrong to put all of the blame on one party. Sylvester had certainly contributed to the delay in publication by making late changes to his own paper and rearranging the order of its contents. However, a letter from C. S. Peirce to Gilman, dated August 7, 1880, included the following comment:[27]

> I have received from Sylvester an account of his difficulty with Story. I have written what I could of a mollifying kind, but it really seems to me that Sylvester's complaint is just. I don't think Story appreciates the greatness of Sylvester, and I think he has undertaken to get the *Journal* into his own control in an unjustifiable degree . . . It is no pleasure to me to intermeddle in any dispute but I feel bound to say that Sylvester has done so much for the University that no one ought to dispute his authority in the management of his department.

By this time, Sylvester was well past what we now think of as normal retirement age. In February 1883, Henry Smith, Oxford University's Savilian Professor of Geometry, died unexpectedly, thereby prompting a search for a successor, preferably an Oxford man. News reached Sylvester, and on March 16 he wrote to Cayley indicating that he would probably offer himself as an applicant, as religious barriers had by then been removed. Sylvester submitted his resignation to Johns Hopkins in the fall of 1883 and returned to Britain on December 21. In January 1884, he wrote to Felix Klein in Germany giving further reasons for leaving America:[28]

> I resigned my position in Baltimore
>
> 1° Because I was anxious to return to my native country
>
> 2° Because I had reasons of a strictly individual and personal nature for wishing to quit Baltimore

3° (and *paramountly*) because I did not consider that my mathematical erudition was sufficiently extensive nor the vigor of my mental constitution adequate to keep me abreast of the continually advancing tide of mathematical progress to that extent which ought to be expected from one on whom practically rests the responsibility of directing and moulding the mathematical education of 55 millions of one of the most intellectual races of men upon the face of the earth.

There has been some discussion as to who really founded the *American Journal of Mathematics*. From the beginning, Gilman had desired all departments of his new university to found research-level journals, and the idea of one in mathematics had independently occurred to Story. But most commentators acknowledge Sylvester as the founder, and at his farewell banquet, on December 20, 1883, Gilman indeed gave him the credit. However, Sylvester's response indicated otherwise:[29]

> You have spoken about our *Mathematical Journal*. Who is the founder? Mr Gilman is continually telling people that I founded it. That is one of my claims to recognition which I strongly deny. I assert that he is the founder. Almost the first day that I landed in Baltimore . . . he began to plague me to found a *Mathematical Journal* on this side of the water—something similar to the *Quarterly Journal of Pure and Applied Mathematics* [of Oxford] . . . Again and again he returned to the charge, and again and again I threw all the cold water I could on the scheme, and nothing but the most obstinate persistence and perseverance brought his views to prevail. To him and to him alone, therefore, is really due whatever importance attaches to the foundation of the *American Journal of Mathematics*.

The reality is that Sylvester had the international standing, with links in Europe and previous experience of being involved in the creation of Oxford's *Quarterly Journal*, of which he was editor until 1878. Independently, Story had formulated the idea of a learned mathematical publication and wanted to be involved in its creation. However, without Gilman's continual encouragement, direction, and belief that such a journal would be of great benefit to mathematics in America, it probably could not have happened as it did in 1878.

What was Sylvester's legacy in the United States? Apart from the *American Journal of Mathematics*, he successfully established at Johns Hopkins University a successful graduate school that invested time and effort into training future researchers. This in turn had an effect on other educational institutions which then established graduate schools, and the level of mathematical research throughout America gradually improved. As a consequence, it was no longer necessary for graduates to journey abroad for postgraduate study, although some continued to do so.

Sylvester was indeed appointed at the age of 69 to the Savilian Chair of Geometry at Oxford University, a position that he held for the rest of his life. In his late 70s, suffering from partial blindness, he returned to London with a deputy appointed to cover his Oxford duties.

An unpredictable, erratic, and flamboyant scholar, Sylvester could be brilliant, quick-tempered, and restless, filled with immense enthusiasm and an insatiable appetite for knowledge. Throughout his life, he had fought for the underdog in society and supported education for the working classes, for women, and for people who were discriminated against. He was awarded many honors and prizes, including his election as a Fellow of the Royal Society in 1839 at the age of 25, and received the Royal Society's Royal Medal in 1861 and the Copley Gold Medal (its highest award) in 1880. The lunar feature Crater Sylvester was named in his honor. He died on March 15, 1897, in London.

As for William Story, Sylvester's departure from Johns Hopkins left him with a similar desire to move to new pastures, and in 1887 he was offered the position of head of mathematics at the newly founded Clark University in Worcester, Massachusetts. His situation is best summed up by Roger Cooke and V. Frederick Rickey:[30]

> There were many reasons why Story might have wanted to leave Hopkins. He was not a full professor there, though he had been there thirteen years. He was not the editor of the *American Journal of Mathematics*, which had been one of his youthful ideas. Finally, he had come to feel that Hopkins was not the wonderful place intellectually that he thought it might and should be . . . But perhaps most importantly of all, he would have the opportunity to develop a department that focused on graduate education and on research. And he could do it the way that he thought best. For all these reasons, it is likely that the opportunity to move to Clark would have attracted Story.

Story did indeed develop a mathematics faculty according to his own ideas—and in particular a doctoral program with twenty-five degrees awarded between 1892 and 1921, nineteen under his direct supervision. Indeed, Story was so successful in his new position that for a time Clark University was considered by some to have the best mathematics department in America. But in spite of all his work, misfortune struck in 1921, when financial problems forced the university to close its graduate

William Story at Clark University.

program, and he was required to resign. In his later years, Story became interested in the history of mathematics and compiled a considerable bibliography of mathematics and mathematicians, which is now in the care of the American Mathematical Society. He died in Worcester, Massachusetts, on April 10, 1930.

C. S. PEIRCE

Charles Sanders Peirce is usually remembered as a philosopher, mathematician, and logician, and for his controversial and unconventional lifestyle. He was born on September 10, 1839, in Cambridge, Massachusetts. As a young boy, he thrived on the intellectual atmosphere prevailing at the family home, where his father, Benjamin Peirce, entertained academics, politicians, poets, scientists, and mathematicians. Although this provided a scholastic environment, his father avoided discipline, fearing that it might inhibit independence of thought. Such an indulgent attitude provided a platform where the younger Peirce could show off

Charles Sanders Peirce
(1839–1914).

his undoubted genius, but it also left him ignorant of how to behave or interact with people. The lack of parental guidance made it difficult for him to fit in to society and led to problems in later life.

C. S. Peirce enrolled at Harvard College at age 15, but he did not shine in his work, preferring to study on his own with books of his own choosing. He graduated with a bachelor's degree in 1859 and entered the Lawrence Scientific School under the influence of his father, where he met with greater success than in his undergraduate years. He received a master's degree from Harvard in 1862 and a bachelor of science degree from the Lawrence Scientific School in 1863, receiving Harvard's first *summa cum laude* degree in chemistry. He remained at Harvard where he carried out graduate research, and in the spring of 1865 he presented the Harvard Lectures on *The Logic of Science*.

From 1859, for nearly thirty years and in parallel with his academic career, Peirce held a position as a part-time assistant at the Coast Survey; some of this time was under his father as director. In 1876, he produced one of his most notable inventions, the *Quincuncial Map Projection*, which was published in the *American Journal of Mathematics* in 1879; this earned him a reputation as one of the great mapmakers of the time. Although his invention was not taken up at the time, it was used in the mid-20th century to display air routes.

Meanwhile, he was producing seminal work in a wide range of subjects, including probability and statistics, psychophysics (or experimental psychology), and species classification. In addition, he carried out major astronomical research and explored mathematical logic, associative algebra, topology, and set theory. But either through choice or because he was considered unsuitable, he failed to obtain a position within a university mathematics department until 1879.

In the early 1860s, C. S. Peirce encountered the four color problem in an *Athenaeum* book review and was probably the first American to take an interest in it. As he later recalled:[31]

> About 1860 De Morgan in the *Athenaeum*, called attention to the fact that this theorem had never been demonstrated; and I soon after offered to a mathematical society at Harvard University a proof of this proposition extending it to other surfaces for which the numbers of colours are greater. My proof was never printed, but Benjamin Peirce, J. E. Oliver, and Chauncey Wright, who were present, discovered no fallacy in it.

His manuscripts, held in the Houghton Library at Harvard University, give no hint of his approach to the problem.

Another manuscript, dated October 1869 and also in the Houghton Library, connects map coloring with his "logic of relatives". Peirce considered that it was

> a reproach to logic and to mathematics that no proof had been found of a proposition so simple.

He also believed that, because of Cayley's interest in logic, Cayley must also have tried such an approach to the four color problem, but had failed. The following comment by Carolyn Eisele refers to Peirce's searches for a solution:[32]

> But his writings over the years are interlaced with references to the problem; his notebooks are full of sketches and diagrams of various regional possibilities reflecting his continuing interest and experimentation. The fragmentary nature of these attempts is evidence of the frustration that never ceased to haunt him.

In the early 1870s, Peirce made a lengthy tour of Europe, and in June of 1870 he visited Augustus De Morgan in London. Although De Morgan was in poor health at the time, it seems likely that their discussions included the four color problem.[33] De Morgan died in 1871.

In 1879, C. S. Peirce was appointed a part-time lecturer in logic in the department of mathematics at Johns Hopkins University, headed by Sylvester. Initially things went well, and Peirce was exposed to new people and ideas. On November 5, 1879, he attended the meeting of the university's scientific association where Story discussed Kempe's paper, and the minutes record that "Remarks were made upon this paper by Mr. C. S. Peirce". At the next meeting, on December 3, Peirce presented

C. S. Peirce at Johns Hopkins University.

a paper on the four color problem; no copy has survived, but the record of the meeting[34] included a comment that Peirce

> discussed a new point in respect to the Geographical Problem of the Four Colors, showing by logical argumentation that a better demonstration of the problem than the one offered by Mr. Kempe is possible.

Peirce's years at Johns Hopkins were perhaps his most productive and significant. He had several papers published in the *American Journal of Mathematics*, with at least one, on map projections, at the request of Sylvester.

But his employment at Johns Hopkins was not to be a long tenure. In October of 1876, Peirce separated from his wife of thirteen years and embarked on a path that would greatly affect his career. He set up home with a woman named Juliette Froissy Pourtalai, divorced his first wife on April 24, 1883, and married Juliette six days later.

Simon Newcomb was a contemporary of Peirce's who had also studied at Harvard under Benjamin Peirce and graduated one year before Charles. Over the years they kept up an active correspondence which displayed a mutual respect, if not a closeness. But in 1884, shortly after

Newcomb was appointed professor of mathematics and astronomy at Johns Hopkins, he felt it his duty to inform the university's trustees that Peirce had been living with his mistress while remaining married to his first wife. The university did not wish to attract scandal by association, so Peirce's contract was not renewed and he was never to hold another academic post.

Thereafter, his only employment and income were from his part-time association with the Coast Survey. He became increasingly quarrelsome and distanced from his superiors, working in isolation at a time when the survey was experiencing a lack of funding. In 1890, he submitted his long-awaited major report to the Survey, which declined to publish it without considerable revision. This Peirce failed to carry out, and at the end of 1891 the Survey ran out of patience and requested his resignation. This left him with no regular income. Much of his work after 1890 was either rejected for publication or remained incomplete.

In later years, Peirce became increasingly withdrawn from public life and colleagues and more erratic in his behavior. As the American mathematician Thomas Scott Fiske later recalled:[35]

> His dramatic manner, his reckless disregard of accuracy in what he termed "unimportant details," his clever newspaper articles describing the meetings of our young Society interested and amused us all . . . He was always hard up, living partly on what he could borrow from friends, and partly on what he got from odd jobs . . . He was equally brilliant, whether under the influence of liquor or otherwise, and his company was prized by the various organizations to which he belonged; and so he was never dropped from any of them even though he was unable to pay his dues.

Indeed, for much of his later life he lived like a social outcast, sometimes even stealing to eat and occasionally having no permanent address. Believing that there was an international plot to undermine and destroy him, his continuing dream was to generate vast wealth from amazing inventions, but this never happened.

Meanwhile, as we saw earlier, William Story continued to be involved with the four color problem. At a meeting of the National Academy of Sciences held in New York City in November of 1899, he presented the paper "The map-coloring problem", but again no manuscript of this has survived.[36] He also corresponded on the subject with C. S. Peirce, and their letters indicate how they, like many others who worked on it, became frustrated about their lack of progress. Peirce's approach to the problem was algebraic, and it is more than likely that they had discussed the subject during their years together (from 1879 to 1884) at Johns Hopkins. A letter from Peirce to Story in August of 1900 mentioned that he had found the proof contained in his unpublished paper, while a double-dated letter from Story to Peirce in December of 1900 shows Story's continued irritation with the problem:[37]

Dec. 1, 1900

> As to my not answering your letter about the four-color problem, I am heartily tired of that subject. I have spent an immense amount of time on it, and all to no purpose. Your first method had occurred to me years ago, but I did not succeed in getting anything out of it.

Dec. 6, 1900

> My delay in sending this off is largely your own fault. You have again reminded me of that fascinating but elusive problem, and I have spent the time since writing the above in trying to solve it, but alas! I believe that the case of exception to Kempe's method requires that the map shall have at least one triangular or quadrilateral district, in which case the pentagon is not the next district to be colored, i.e. the exception does not occur. But I cannot prove it.

This extract indicates that Story had received some communication from Peirce between the two dates of the letter, suggesting a further approach that Story had attempted without success.

Although Charles Sanders Peirce was not accorded the undoubted recognition that he deserved during his lifetime, there is now a growing interest in his work, especially in logic. Some believe that he was the greatest inventive intellect to have been born on the American continent, and in 1959 Bertrand Russell wrote:[38]

> Beyond doubt . . . he was one of the most original minds of the late nineteenth century and certainly one of the greatest American thinkers ever.

Peirce died of cancer on April 19, 1914, in isolation on his farm in Milford, Pennsylvania.

* * * * *

The efforts of scholars such as Sylvester, Story, and Peirce provided a climate for American-born and American-educated mathematicians to make significant research contributions to mathematics in general, and to the study of trees and the four color problem in particular.

Additionally, the appearance of graduate schools—Johns Hopkins at first, and then Harvard, Yale, Princeton, and others—advanced the development of American mathematics and postgraduate study; this continued through the 1890s and into the 20th century. By 1910, several professors had received all their training in America and would go on to earn international reputations. This healthy state of affairs would pave the way for the United States to become the leading country for mathematics by the middle of the 20th century.

The solid foundations that had been laid, and the encouragement that had been given by the leading mathematicians around the turn of the century, provided a foundation for American mathematicians to make their mark. In the early 20th century, American mathematicians were beginning to take an interest in graph theory and the four color problem, a move that would be boosted by the approaches of Oswald Veblen and George Birkhoff, as we see in Chapter 2. But first, we report on some related progress that was taking place around Europe.

Interlude A
Graph Theory in Europe 1

As American mathematicians began to take an active interest in graph theory in the last quarter of the 19th century, developments were continuing in Europe. In Scotland, P. G. Tait tried to simplify Kempe's proof of the four color theorem by coloring the boundary edges of a map rather than the countries. In England, Percy Heawood pointed out the flaw in Kempe's proof and investigated colorings of maps on surfaces other than the sphere. Later explorations into coloring maps on surfaces were made by Lothar Heffter of Germany, who studied orientable surfaces, and Heinrich Tietze of Austria, who investigated non-orientable ones. In Denmark, Julius Petersen discussed the factorization of graphs, while in Germany, Hermann Minkowski tried to convince his students that he could prove the four color theorem.

These writers used differing terminologies when describing maps. Countries (called "districts" by Kempe and Story) were often called "regions", whereas the boundaries of a country were called "lines", "edges", or "sides", and meeting points of countries and boundaries were "points" or "vertices". We shall use these terms interchangeably, generally following the use by the person whose results are being discussed.

P. G. TAIT (SCOTLAND)

Around 1880, Peter Guthrie Tait, a distinguished applied mathematician from Edinburgh, learned of Kempe's proof of the four color problem, which was accepted as correct at the time. Convinced that he could greatly simplify it, he presented some shorter proofs to the Royal Society of Edinburgh,[1] but these were also deficient.

However, his attempts contained one useful and original idea. Instead of coloring the countries of a cubic map, he colored their boundaries, obtaining the following result:

P. G. Tait (1831–1901).

> The countries of a cubic map can be colored with four colors if and only if their boundaries can be colored with three colors, with adjacent boundaries colored differently.

He claimed that a proof of his equivalent version "is given easily by induction"—but he was mistaken. His version was as difficult to prove as the original four color theorem.

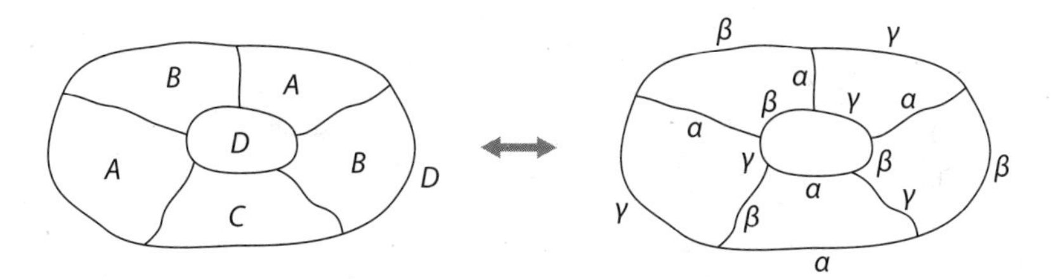

Coloring the countries and boundaries of a cubic map.

Four years later, Tait returned to coloring the boundaries and proved that if a cubic map has a *Hamiltonian cycle*—that is, a closed cycle that passes through every meeting point—then one can indeed color its boundaries with three colors and its countries with four colors.[2] He then claimed that every cubic polyhedron has this Hamiltonian property—a statement that he was unable to prove. Indeed, the English mathematician Thomas Kirkman was unable to decide whether Tait's claim was actually true: [3]

> The theorem—that on every *p*-edron *P*, having only triedral summits, a closed circle of edges passes once through every summit, has this provoking interest, that it mocks alike at doubt and proof.

Eventually, in 1946, the English mathematician W. T. Tutte showed Tait's claim to be false by exhibiting a counter-example with 46 points (see Chapter 5).

PERCY HEAWOOD (ENGLAND)

Percy John Heawood studied at Oxford University, where he learned about the four color problem from Henry Smith, the Savilian Professor of Geometry. In 1887, he was appointed a lecturer at the Durham Colleges, later to become the University of Durham, where he had a long and distinguished career.

In 1890, Heawood wrote a groundbreaking paper, "Map-colour theorem",[4] in which he exposed the fundamental flaw in Kempe's accepted proof of the four color theorem, giving a specific example that showed why Kempe was mistaken in trying to perform two simultaneous color interchanges for the pentagon. By modifying Kempe's arguments, Heawood was able to prove that the countries of every map on the plane or sphere can be colored with five colors—itself a remarkable result. Unfortunately, Heawood's revelation of Kempe's error remained largely

Percy John Heawood (1861–1955).

unknown for several years, and he received little recognition for his achievement.[5]

Eight years later, Heawood wrote a second paper that developed Tait's idea of coloring the boundaries of a cubic map with three colors.[6] He first showed that if the number of boundaries around every country is divisible by 3, then the countries can be colored with four colors. He then explained:

> If we can assign the numbers 1 and –1 to the meeting points of a cubic map so that the sum of the numbers around each country is divisible by 3, then we can color the boundary lines with three colors, and the countries with four colors.

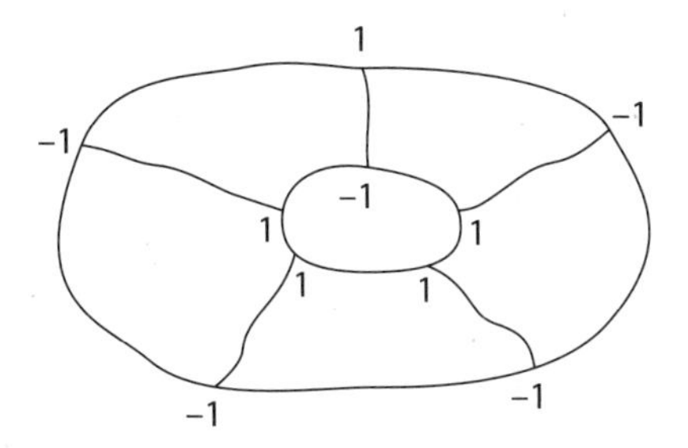

Labeling the points of a cubic map.

Moreover, if the points are labeled $v_1, v_2, \ldots, v_n$, then a system of congruences of the form

$$x_i + x_j + \cdots + x_k \equiv 0 \pmod{3}$$

can be generated, with one congruence for each country. Here, each unknown x_i is either 1 or –1, and x_i appears in the congruence for a particular country if and only if the point v_i lies on a boundary of that country.

Heawood's 1890 paper also discussed the coloring of cubic maps on surfaces other than a plane or sphere, such as a torus, or the surface of a sphere to which several handles have been added (or, equivalently, of a many-holed donut). Such a surface is called an *orientable surface*. It has *genus g* if there are g handles or holes, and is denoted by S_g; for example, the torus S_1 has genus 1 and the sphere S_0 has genus 0.

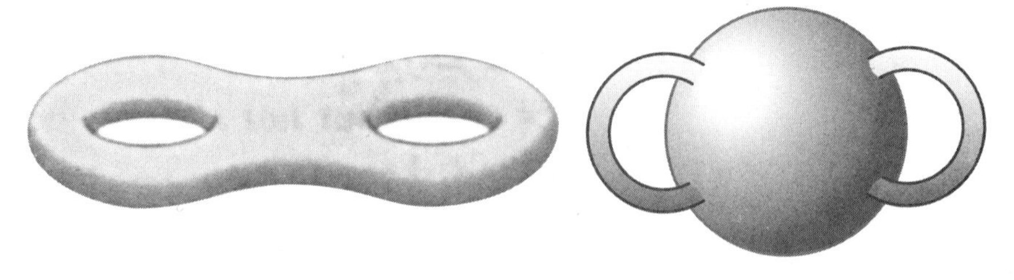

Two drawings of the surface S_2 of genus 2.

The *chromatic number* $\chi(S)$ of a surface S is the smallest number of colors needed to color the regions of every map on S, with neighboring regions colored differently. Heawood showed that the chromatic number $\chi(S_1)$ of a torus is 7—that is, every map on a torus can be colored with seven colors, and there are torus maps that require all seven colors.

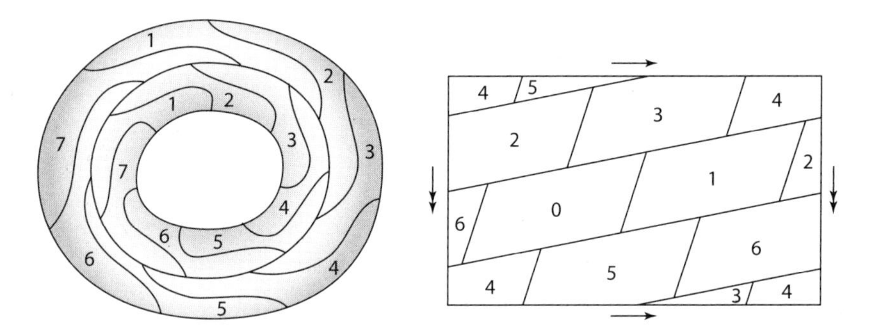

Two drawings of a torus map that requires seven colors; in the right-hand drawing the opposite sides of the rectangle are to be identified as indicated.

In trying to establish the chromatic number of the surface S_g, where $g \geq 1$, Heawood used an extension of Euler's formula (due to the Swiss mathematician Simon-Antoine-Jean L'Huilier[7]) for maps drawn on it—namely, that if such a map has R regions, B boundaries, and P points, then

$$R + P = B + 2 - 2g.$$

He deduced that, for $g \geq 1$, the countries of every cubic map on S_g can be colored with $H(g)$ colors, where

$$H(g) = \left\lfloor \tfrac{1}{2}\left(7 + \sqrt{1 + 48g}\right) \right\rfloor.$$

Here the outer brackets represent the "floor function", which indicates that the result must be rounded down when it is not a whole number. So for the torus S_1 and the two-holed torus S_2,

$$H(1) = \lfloor \tfrac{1}{2}(7+\sqrt{49}) \rfloor = \lfloor 7 \rfloor = 7 \quad \text{and} \quad H(2) = \lfloor \tfrac{1}{2}(7+\sqrt{97}) \rfloor = \lfloor 8.4244\ldots \rfloor = 8.$$

It follows that

$$\chi(S_g) \le \lfloor \tfrac{1}{2}(7+\sqrt{1+48g}) \rfloor.$$

But Heawood did not consider it necessary to prove that, apart from the torus, there are maps on S_g that actually require the number of colors predicted by this formula, so that the inequality becomes an equality. This major omission later came to be known as the *Heawood conjecture*, which can be stated as follows:

> *Heawood conjecture*: For each $g \ge 1$, the chromatic number of the orientable surface S_g is $\chi(S_g) = \lfloor \tfrac{1}{2}(7+\sqrt{1+48g}) \rfloor$.

It would be over seventy years before this was proved in general (see Chapter 6).

Percy Heawood is remembered mainly for exposing Kempe's error and for the Heawood conjecture, but he never lost interest in the four color problem. He continued to write papers on the congruences associated with map coloring until his 90th year.[8]

JULIUS PETERSEN (DENMARK)

Julius Peter Christian Petersen was a Danish mathematician widely known for his successful school and undergraduate textbooks, as well as for his research into geometry. In 1891, he published a pioneering paper, "Die Theorie der regulären Graphs" (The theory of regular graphs),[9] on factorizing graphs. His choice of the English word "Graphs", rather than the German "Graphen", arose from his extensive discussions on the subject with Sylvester.[10]

A *regular graph* is a graph in which every vertex has the same degree, and an *r-factor* in a graph is a subgraph that is regular of degree r and includes every vertex. In particular, a 1-factor—sometimes called a *perfect*

Julius Petersen (1839–1910).

or *complete matching*—is a collection of disjoint edges that includes every vertex of G.

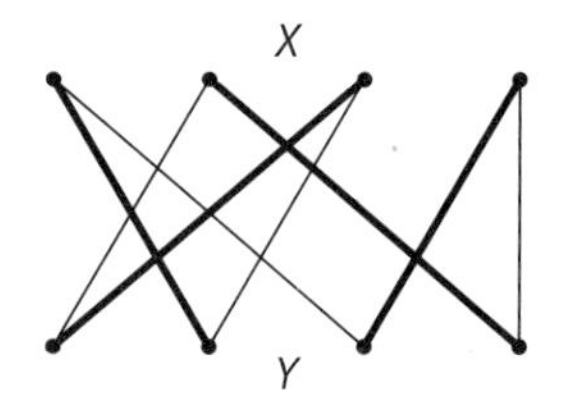

A 1-factor in a regular graph, matching the vertices in the set X with those in the set Y.

Further, an *r-factorization* of G splits G into r-factors; for example, the complete graph K_7, which is regular of degree 6, has a 2-factorization into three 2-factors.

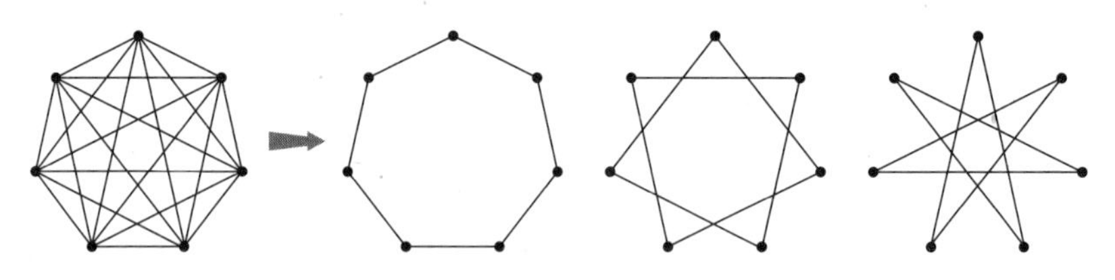

A 2-factorization of K_7.

Petersen asked when it is possible to split a regular graph of degree k into regular factors of given degree r, and proved that

If k is even, then any graph that is regular of degree k can be split into 2-factors.

He also observed that the situation is more complicated when k is odd.

Petersen was particularly interested in factors in cubic graphs, where $k=3$. His main achievement was to prove the following:

Petersen's theorem: Every cubic graph with at most two leaves has a 1-factor.

Here, Petersen thought of a *leaf* as a part of the graph that is separated from the rest by the removal of a single edge. (Nowadays, the word "leaf" refers to just a single vertex of degree 1.) Sylvester's graph is a cubic graph with three leaves and no 1-factor, so Petersen's condition on leaves is necessary.

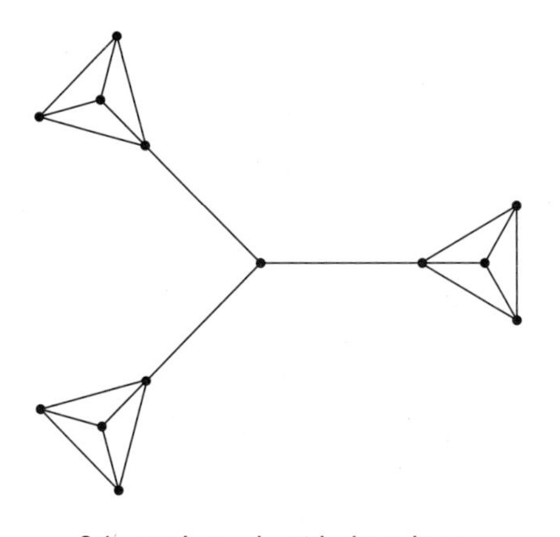

Sylvester's graph, with three leaves and no 1-factor.

Petersen's proof of his theorem was complicated, and simpler proofs were provided later by the Americans H. Roy Brahana and Orrin Frink, and by the Belgian Alfred Errera.[11]

We have seen that Tait proved that the four color theorem is true for cubic maps if and only if the boundaries can be colored with three colors with all three colors appearing at each point. Such a coloring of the edges gives a 1-factorization, and Tait conjectured that every cubic graph without leaves has such a 1-factorization. In 1898, Petersen wrote a short note in which he disproved Tait's conjecture by constructing a cubic graph with no 1-factorization.[12] This graph is now called the *Petersen graph*, although Kempe had already published a drawing of it in 1886.[13]

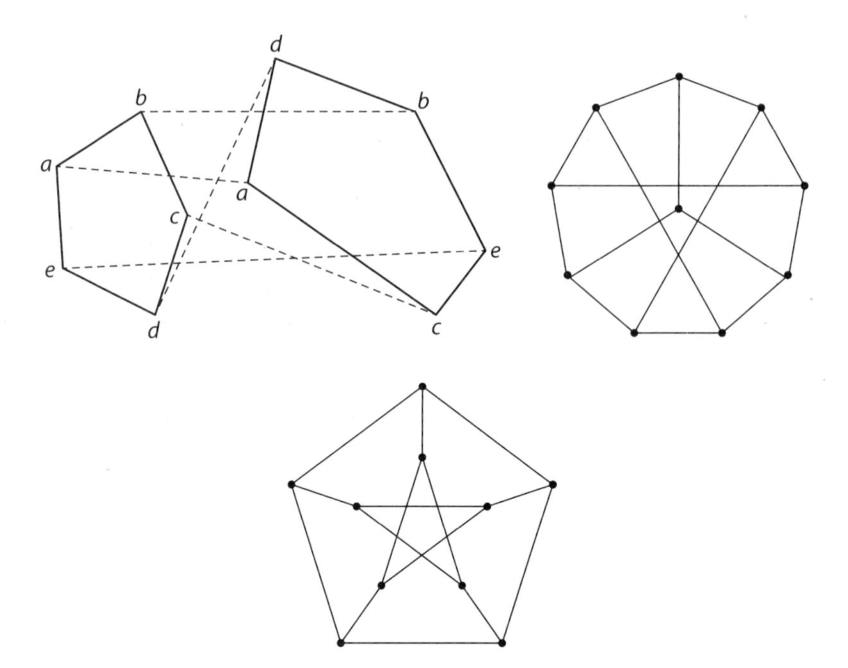

Three drawings of the Petersen graph.

In a later note on connections between the four color problem and the factorization of graphs,[14] Petersen commented that Kempe had

> only skimmed over the problem and committed his error just where the difficulties began

and concluded by remarking, somewhat unexpectedly, that

> I am not certain of anything, but if I had to wager, I would hold that the theorem of four colors is not correct.

LOTHAR HEFFTER (GERMANY)

Lothar Wilhelm Julius Heffter was a German mathematician who studied and lectured at several universities. In a paper of 1891,[15] written while teaching in Giessen, he extended Heawood's results on the coloring of maps on the orientable surface S_g.

It was Heffter who spotted the gap in Heawood's paper. Accepting Heawood's proof that $H(g) = \left\lfloor \tfrac{1}{2}\left(7 + \sqrt{1 + 48g}\right) \right\rfloor$ colors are sufficient for coloring maps on the orientable surface S_g—that is, $\chi(S_g) \leq$

Lothar Heffter (1862–1962).

$\lfloor \frac{1}{2}(7+\sqrt{1+48g}) \rfloor$—he pointed out that Heawood had neglected to show that when $g \geq 2$, there are maps that require this number of colors, so that the inequality becomes an equality. Heffter was determined to fill this gap, and managed to do so for all values of g up to 6 and for a few other values, but he was unable to produce an argument that worked in general.

His approach was first to investigate maps in which every region (country) meets every other one, calling this a *system of neighboring regions*; for example, there are up to seven neighboring regions on a torus, as we have seen. We note that, if there are n neighboring regions, then n colors are needed to color them.

Heffter then turned the problem inside out! Instead of fixing the genus g and asking for the largest number n of neighboring regions that can be drawn on the surface S_g, he fixed the number n of neighboring regions and asked for the smallest genus g for which n neighboring regions can be drawn on S_g; for example, if there are seven neighboring regions, then the smallest value of g is 1 (the torus). Using L'Huilier's generalized version of Euler's formula, Heffter obtained the lower bound

$$g \geq \lceil \tfrac{1}{12}(n-3)(n-4) \rceil.$$

Here, the outer brackets represent the "ceiling function", which indicates that the result must be rounded up when it is not a whole number; for example, when $n=7$, it is

$$\lceil \tfrac{1}{12}(7-3)(7-4) \rceil = \lceil 1 \rceil = 1,$$

as expected. With this formulation, the Heawood conjecture takes the following form:

Heawood conjecture (neighboring regions): The simplest orientable surface on which n (≥ 3) neighboring regions can be drawn is S_g, where $g = \lceil \tfrac{1}{12}(n-3)(n-4) \rceil$.

Heffter was able to prove this for all values of n up to 12 and for a few other values.

Heffter's final inversion of the problem involved the idea of duality, as introduced by Kempe (see Chapter 1). By replacing each of the n neighboring regions by a point inside it, and joining neighboring points, we obtain a system of n interconnected points—that is, the complete graph K_n.

We now define the *orientable genus* $g(G)$ of a graph G to be the smallest value of g for which G can be drawn without crossings on the orientable surface S_g; for example, the orientable genus of the complete graph K_7 is 1 (the torus).

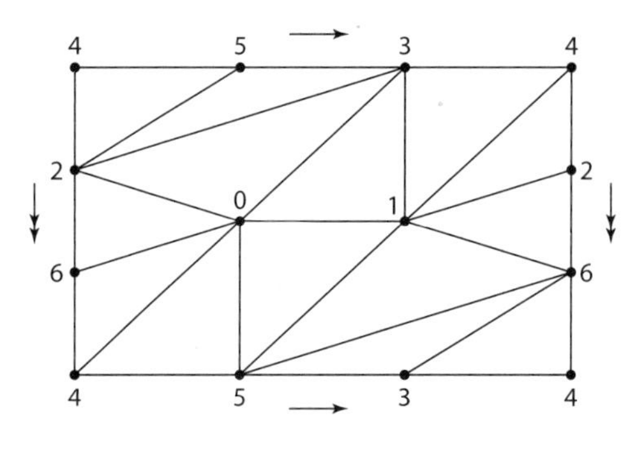

A drawing of K_7 on a torus.

The Heawood conjecture then takes the following form:

Heawood conjecture (complete graphs): The orientable genus of K_n ($n \geq 3$) is

$$g(K_n) = \lceil \tfrac{1}{12}(n-3)(n-4) \rceil.$$

This was later called the *thread problem*, because one can think of connecting the n vertices of K_n by a simple curve (a thread) on the surface so that these curves do not intersect.[16]

It turns out that these three equivalent versions of the Heawood conjecture are true for all orientable surfaces, as we shall see in Chapter 6. Moreover, although Heffter was unable to prove the conjecture in general, it is significant that his formula for the orientable genus included the number 12, as the eventual proof split into twelve separate cases, depending on the remainder when n is divided by 12.

HEINRICH TIETZE (AUSTRIA)

Almost twenty years later, the Austrian mathematician Heinrich Franz Friedrich Tietze was thinking along similar lines. He would later become known as the author of the popular *Famous Problems of Mathematics*, which includes some attractive color plates of maps on surfaces.[17]

Whereas Heawood and Heffter had been concerned with maps drawn on orientable surfaces (spheres with handles added), Tietze was interested in maps on *non-orientable surfaces*. These are spheres with "cross-caps" added, where a *cross-cap* is obtained by cutting a hole in a sphere and identifying its boundary with the edge of a Möbius band. These surfaces include the projective plane and the Klein bottle (first described in 1882 by the German mathematician Felix Klein), which cannot be embedded in three-dimensional space.

Heinrich Tietze (1880–1964).

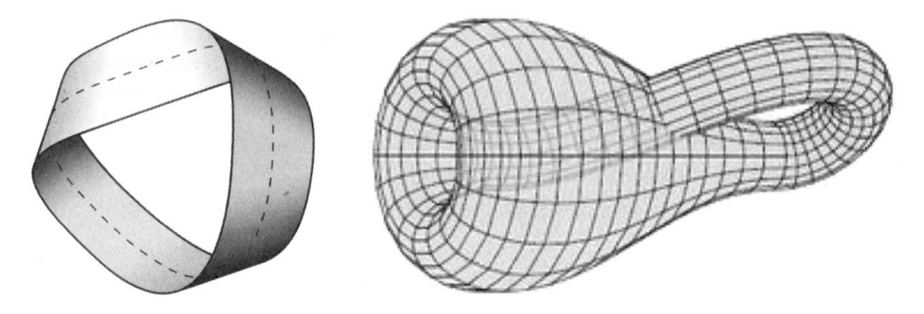

A Möbius band and a Klein bottle.

Just as a torus can be drawn as a rectangle with opposite sides appropriately identified, a Möbius band, a projective plane, and a Klein bottle can each be similarly represented.

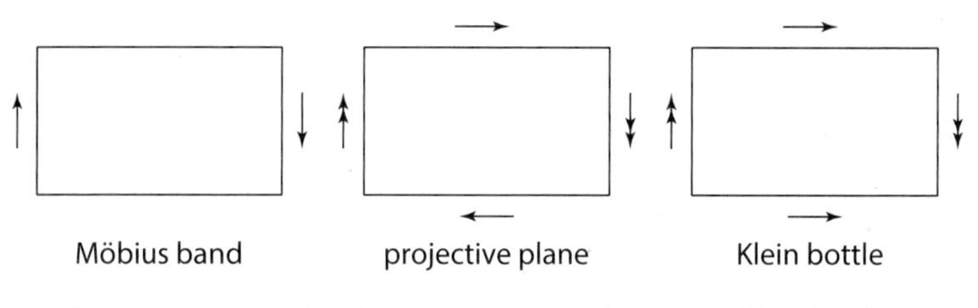

Representing a Möbius band, a projective plane, and a Klein bottle.

A non-orientable surface has *genus q* if it is obtained by adding q cross-caps to a sphere, and is denoted by N_q. In particular, the projective plane N_1 has genus 1, and the Klein bottle N_2 has genus 2.

In 1910, Tietze wrote a paper in which he showed that up to six neighboring regions can be drawn on a Möbius band.[18]

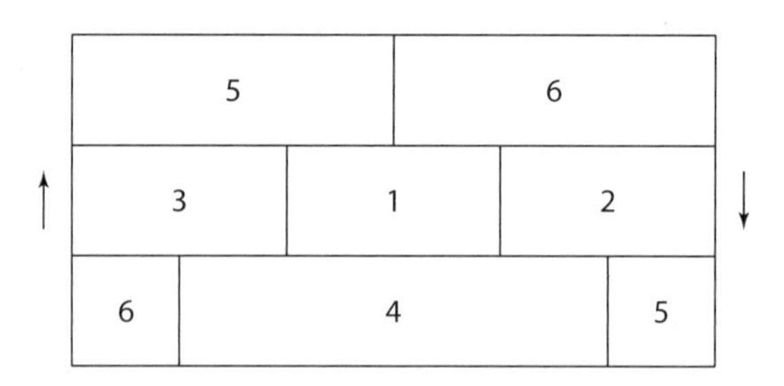

Six neighboring regions on a Möbius band.

He then imitated the arguments of Heawood and Heffter in trying to establish the chromatic number $\chi(\mathcal{N}_q)$ of the non-orientable surface $\mathcal{N}_q$, where $q \geq 1$. To this end, he used an extension of Euler's formula for maps drawn on $\mathcal{N}_q$—namely, that if such a map has R regions, B boundaries, and P points, then

$$R + P = B + 2 - q.$$

He deduced that, for $q \geq 1$, the regions of every cubic map on $\mathcal{N}_q$ can be colored with $\mathcal{T}(q)$ colors, where

$$\mathcal{T}(q) = \left\lfloor \tfrac{1}{2}\left(7 + \sqrt{1 + 24q}\right) \right\rfloor,$$

so for the projective plane $\mathcal{N}_1$ and the Klein bottle $\mathcal{N}_2$,

$$\mathcal{T}(1) = \left\lfloor \tfrac{1}{2}(7 + \sqrt{25}) \right\rfloor = \lfloor 6 \rfloor = 6 \quad \text{and} \quad \mathcal{T}(2) = \left\lfloor \tfrac{1}{2}(7 + \sqrt{49}) \right\rfloor = \lfloor 7 \rfloor = 7.$$

It follows that

$$\chi(\mathcal{N}_q) \leq \left\lfloor \tfrac{1}{2}\left(7 + \sqrt{1 + 24q}\right) \right\rfloor.$$

Again, the difficulty is to prove that there are maps on $\mathcal{N}_q$ that actually require the number of colors predicted by this formula, so that the inequality becomes an equality. The Heawood conjecture for non-orientable surfaces then takes the following form:

> *Heawood conjecture* (*map coloring*): For each $q \geq 1$, the chromatic number of the non-orientable surface $\mathcal{N}_q$ is $\chi(\mathcal{N}_q) = \left\lfloor \tfrac{1}{2}\left(7 + \sqrt{1 + 24q}\right) \right\rfloor$.

But here Tietze ran into a difficulty. He proved that every map on a Klein bottle (where $q = 2$) can be colored with seven colors, and he found a Klein bottle map that requires six colors, but he was unable to find one that requires the predicted seven colors. Whether such maps exist was not resolved until the 1930s (see Chapter 3).

Tietze then turned the problem inside out, as Heffter had done, by fixing the number n of neighboring regions, and asking for the smallest genus q for which n neighboring regions can be drawn on the surface $\mathcal{N}_q$. He obtained the lower bound

$$q \geq \left\lceil \tfrac{1}{6}(n - 3)(n - 4) \right\rceil;$$

for example, when $n=6$ it is

$$\lceil \tfrac{1}{6}(6-3)(6-4) \rceil = \lceil 1 \rceil = 1.$$

With this formulation, the Heawood conjecture for non-orientable surfaces then takes the following form:

Heawood conjecture (*neighboring regions*): The simplest non-orientable surface on which n (≥ 3) neighboring regions can be drawn is N_q, where $q = \lceil \tfrac{1}{6}(n-3)(n-4) \rceil$.

We now define the *non-orientable genus* $\hat{g}(G)$ of a graph G to be the smallest value of q for which G can be drawn without crossings on the non-orientable surface N_q; for example, the non-orientable genus of the complete graph K_6 is 1 (corresponding to the surface N_1, the projective plane).

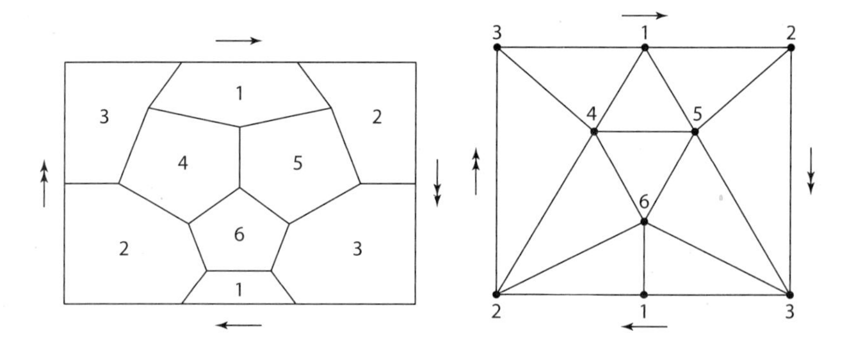

Six neighboring regions and a drawing of K_6 on a projective plane.

The Heawood conjecture then takes the following form:

Heawood conjecture (*complete graphs*): The non-orientable genus of K_n ($n \geq 3$) is

$$\hat{g}(K_n) = \lceil \tfrac{1}{6}(n-3)(n-4) \rceil.$$

It turns out that these equivalent versions of the Heawood conjecture are true for all non-orientable surfaces, *except one*, as we shall discover in Chapters 3 and 6.

HERMANN MINKOWSKI (GERMANY)

Hermann Minkowski was a Lithuanian-born naturalized German mathematician who taught at Göttingen University from 1902 until his early death in 1909. He is remembered for his contributions to number theory, for originating the geometry of numbers, and for his work on mathematical physics and the theory of relativity.

Hermann Minkowski (1864–1909).

But, for all his brilliance, Minkowski was unable to prove the four color theorem. It is recorded that he interrupted a lecture he was giving on analysis situs (topology) to tell his student audience about it. The story is as follows:[19]

> "This theorem has not yet been proved, but that is because only mathematicians of the third rank have occupied themselves with it," Minkowski announced to the class in a rare burst of arrogance. "I believe I can prove it."
>
> He began to work out his demonstration on the spot. By the end of the hour he had not finished. The project was carried over to the next meeting of the class. Several weeks passed in this way. Finally, one rainy morning, Minkowski entered the lecture hall, followed by a crash of thunder. At the rostrum, he turned towards the class, a deeply serious expression on his face.
>
> "Heaven is angered by my arrogance," he announced. "My proof of the Four Color Theorem is also defective."
>
> He then took up the lecture on topology at the point where he had dropped it several weeks before.

* * * * *

Although many mathematicians devoted significant time and effort to tackling the four color problem, no solution had been found. Europeans had carried out most of this work, but this situation would soon change as American mathematicians began to contribute to the subject. The next chapter introduces three of these: Paul Wernicke, Oswald Veblen, and George Birkhoff.

Chapter 2
The 1900s and 1910s

Oswald Veblen and George Birkhoff, two of America's most important mathematicians in the early 20th century, made significant contributions to the development of graph theory. Between their first meeting in Chicago in 1902 and the latter's death in 1944, they remained close friends and colleagues in their mutual desire to advance and promote mathematics in the United States.

Birkhoff's first significant contribution to the subject appeared in 1912, in the same volume of the *Annals of Mathematics* as a paper by Veblen. These two papers, a follow-up paper by Birkhoff, and some lectures and a book by Veblen, were to provide the impetus for the ever-growing American interest in graph theory, and in map coloring in particular.

We also look here at the involvement of some American mathematicians in World War I. But first we examine the notable contribution of another American who has remained largely unknown within the mathematical community.

PAUL WERNICKE

Paul August Ludwig Wernicke was born in Berlin in 1866. He migrated to the United States in 1893 and by 1900 was already a naturalized American citizen. From 1894 to 1906, he was a professor of modern languages at the State College of Kentucky, but during this period his interests turned increasingly toward mathematics and he was soon elected a member of the American Mathematical Society.

As with many others at the time, Wernicke became captured by the four color problem. His first contribution to mathematics was to present a talk on this topic at the American Mathematical Society's summer meeting in Toronto in 1897, under the chairmanship of the society's president, Simon Newcomb. An abstract of his talk appeared in the society's *Bulletin*:[1]

> Given a map correctly colored and with its frontiers marked, the author proves that any triangles, quadrangles, and pentagons can be introduced and correctly marked at the same time. The main theorem then follows by induction.

Based on Tait's connection between colorings of the boundaries and the countries of a cubic map, Wernicke's approach seems to have involved adding new countries to a map to convert it into one that he could color with four colors, but nothing appears to have come from it.

For a while, Wernicke spent some time back in Germany, studying at the University of Göttingen with Hermann Minkowski, and he received his doctoral degree in mathematics in 1903 for the thesis *Über die Analysis situs mehrdimensionaler Räume* (On the Analysis of Position for Higher-dimensional Spaces). He then returned to the United States, and in 1904 his best-known paper, which he had written while in Göttingen, made its appearance.[2]

Kempe had shown that every cubic map must contain a digon, triangle, square, or pentagon. We call such a set of regions an *unavoidable set*, in the sense that every cubic map must contain at least one of them. Wernicke's achievement was to extend Kempe's result by showing that if such a map contains no digon, triangle, or square, then not only must it include a pentagon, but there must be a pentagon adjacent to another pentagon, or a pentagon adjacent to a hexagon. His hope was that these might prove more amenable to analysis than the single pentagon that had defeated Kempe. Wernicke's new unavoidable set was the most significant contribution to solving the four color problem since those of Kempe and Heawood.

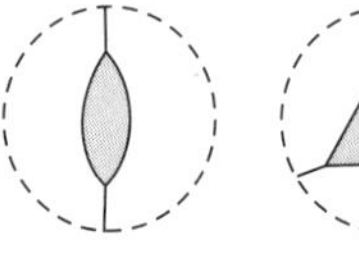

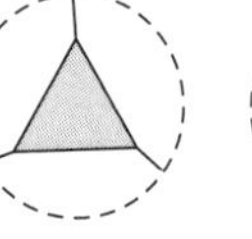

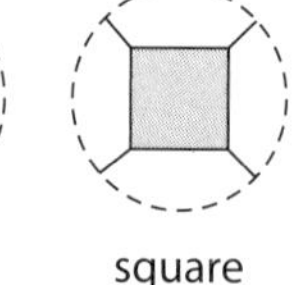

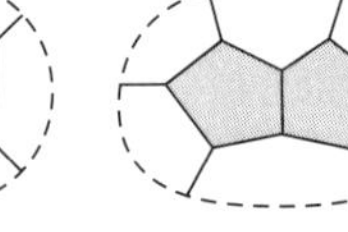

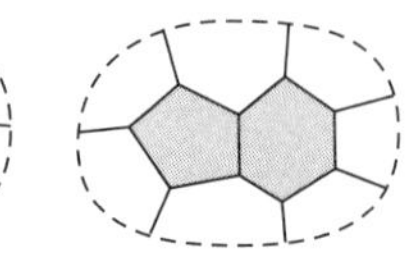

Wernicke's unavoidable set.

OSWALD VEBLEN

Oswald Veblen's paternal grandparents emigrated from Norway to the United States in 1847, where one of their children, Oswald's father, subsequently became a professor of mathematics and physics at the University of Iowa. Oswald was born in Decorah, Iowa, in 1880, and entered that university at the age of 14, graduating with a bachelor's degree in 1898. After spending another year there as an assistant in the physics department, he transferred to Harvard University for a year, earning a second bachelor's degree in 1900.

On leaving Harvard, Veblen spent three years of graduate study at the University of Chicago, gaining much of his early mathematical inspiration from his supervisor, E. H. Moore, and from Moore's German colleagues, Oskar Bolza and Heinrich Maschke, who were also on the teaching staff. As a fellow student recalled:[3]

> Moore was brilliant and aggressive in his scholarship, Bolza rapid and thorough, and Maschke more brilliant, sagacious and without doubt one of the most delightful lecturers on geometry of all times.

Veblen's doctoral dissertation on *A System of Axioms for Geometry* was partly inspired by the work of Henri Poincaré in France and earned him his degree in 1903.

Oswald Veblen
(1880–1960).

From 1905 to 1932, Veblen taught mathematics at Princeton University, initially as an instructor and then as a full professor from 1910 onward. During the academic year 1928–29 he taught at Oxford University in England, as part of an exchange arrangement with G. H. Hardy, after which he became the first professor of mathematics at the newly established Institute for Advanced Study in Princeton.

Oswald Veblen's research and influence ranged over many areas of mathematics, including symbolic logic and the foundations of geometry and topology. Through his work, and that of his students, Princeton became one of the leading centers in topology, earning Veblen the designation of "Statesman of mathematics".[4]

Veblen's two contributions to graph theory were influential and for many years provided among the best introductions to the subject. It can only be speculated as to how he became interested in this area, but during his time in Iowa and Chicago, he may have attended mathematical meetings that presented papers on the subject, such as those of C. S. Peirce or Paul Wernicke. Additionally, he would have had access to learned journals and may have read the contributions of Kempe, Heawood, and others. There seems to be no evidence of Veblen's having had any direct contact with either Peirce or Wernicke.

Modular Equations

When Veblen's paper on the four color problem was published in 1912, topology was still not widely pursued in America. The first of his writings on graph theory was the following paper,[5] presented to the American Mathematical Society on April 27, 1912. In it, Veblen drew on ideas from finite geometry and incidence matrices, with the adjacencies between the vertices (meeting points), edges (boundaries), and countries of a map providing the matrix entries. He expressed various results on maps, including the four color problem, in an algebraic form. His use of matrices followed that of Henri Poincaré.[6]

O. Veblen: *An application of modular equations in analysis situs* (1912)

Veblen began by showing that any map yields two matrices, remarking that:

> These matrices are identical on interchanging rows and columns with those employed by Poincaré if the + and − signs used by the latter are omitted.

He motivated his analysis with the complete graph K_4, which he described as

> the map obtained by projecting an inscribed tetrahedron from one of its interior points to the surface of a sphere.

Matrix A takes the four vertices V_1 to V_4 of K_4 as rows, and the six edges x_1 to x_6 as columns, with 1 appearing when the vertex meets the edge, and 0 otherwise. Matrix B takes the edges of K_4 as rows and the countries C_1, C_2, C_3, and C_4 as columns, with 1 appearing when the edge borders the country and 0 otherwise.

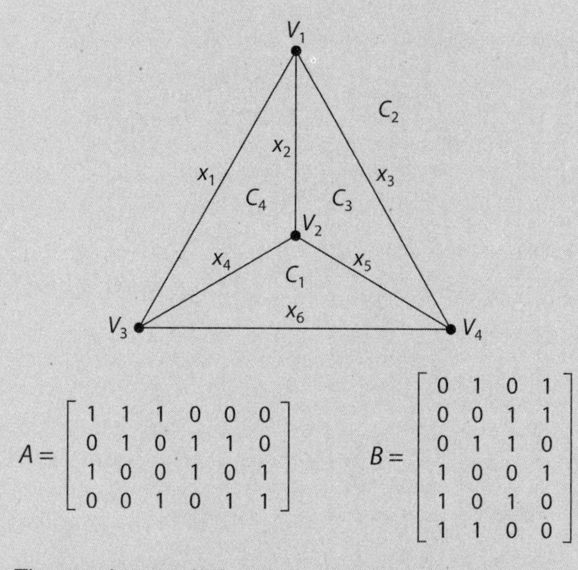

$$A = \begin{bmatrix} 1 & 1 & 1 & 0 & 0 & 0 \\ 0 & 1 & 0 & 1 & 1 & 0 \\ 1 & 0 & 0 & 1 & 0 & 1 \\ 0 & 0 & 1 & 0 & 1 & 1 \end{bmatrix} \qquad B = \begin{bmatrix} 0 & 1 & 0 & 1 \\ 0 & 0 & 1 & 1 \\ 0 & 1 & 1 & 0 \\ 1 & 0 & 0 & 1 \\ 1 & 0 & 1 & 0 \\ 1 & 1 & 0 & 0 \end{bmatrix}$$

The complete graph K_4, and the incidence matrices A and B.

From each matrix, Veblen developed two systems of linear equations, with all calculations carried out (modulo 2), so that $1+1=0$. For matrix A, the first system has a variable x_a for each edge a, and an equation of the form

$$x_a + x_b + x_c + \cdots = 0$$

for each vertex; for example, the equation arising from the first row of matrix A is $x_1+x_2+x_3=0$. Solving these equations yields a way of labeling the edges with 0s and 1s, such that the number of 1s labeling the edges at each vertex is even. As Veblen observed:

> The edges labeled with 1's in this manner form a number of closed circuits no two of which have an edge in common.

He deduced that the number of linearly independent solutions is $\alpha_2 - 1$, where α_2 is the number of countries, and he called them "fundamental solutions".

The second system of equations for the matrix A has a variable V_a for each vertex, and an equation of the form

$$V_a + V_b = 0$$

for each edge V_aV_b. Veblen proved that the rank of the matrix A is $\alpha_0 - 1$, and on combining this with the above result, he deduced that

$$\alpha_1 - (\alpha_0 - 1) = \alpha_2 - 1, \quad \text{or} \quad \alpha_0 - \alpha_1 + \alpha_2 = 2,$$

which is Euler's formula.

For matrix B, the first system of equations has a variable y_a for each country, and an equation of the form

$$y_a + y_b = 0$$

for the countries meeting along each edge. The second system has a variable e_a for each edge, and an equation of the form

$$e_a + e_b + \cdots = 0$$

for the edges surrounding each country. Again, he was able to deduce Euler's equation.

Having established his four systems of linear equations, Veblen turned to the four color problem. Using the four elements, 0, 1, i, and $i+1$, to represent the colors, where $i^2 + i + 1 = 0$ and the calculations are carried out (modulo 2), he observed that a solution to the four color problem consists in finding a set of values $(y_1, y_2, \ldots, y_{\alpha_2})$ that satisfies none of the above equations $y_a + y_b = 0$.

Veblen continued by remarking that the set of non-zero values $(y_1, y_2, \ldots, y_{\alpha_2})$ can be regarded as a point in a finite projective space of $\alpha_2 - 1$ dimensions. He then explored different subspaces of this projective space, before reformulating the four color problem in terms of them. More precisely, in a finite projective space of $\alpha_2 - 1$ dimensions with three points on each line, there are a number of subspaces $S^C_{\alpha_2 - n}$ with dimension $\alpha_2 - n$, one for each odd cycle C_n, and these subspaces all have one point in common. Veblen then asserted that

> The map can be colored in four colors if and only if there exists a point not on any of these $S^C_{\alpha_2 - n}$'s. There are as many distinct ways of coloring the map (aside from permutations of the colors) as there are real lines in the $(\alpha_2 - 1)$-space which do not meet any $S^C_{\alpha_2 - n}$ (n, odd).

To conclude, Veblen investigated those solutions of the equations that provide conditions under which the four color problem can be solved, pointing out that these equations are essentially the same as the congruence equations derived by Heawood in his paper of 1898 (see Interlude A). The final paragraph, which sums up the paper's content, reads:

> To solve the four color problem it is necessary and sufficient to find a solution of these equations in which none of the variables vanish. The variables may be interpreted as coördinates of points in a finite projective space of α_0-dimensions in which there are four points on every line.

Veblen's Colloquium Lectures

Every few years, the American Mathematical Society selected two prominent mathematicians to give a series of summer Colloquium Lectures to its membership—one of the highest recognitions that a mathematician in America could receive. These lectures were always well attended, giving rise to considerable discussion, and were then revised, extended, and collected together as a monograph.

In 1916, Oswald Veblen presented the Colloquium Lectures in Cambridge, Massachusetts, but because of World War I and the society's lack of funds, their publication was delayed. The monograph eventually appeared in 1922 under the title of *Analysis Situs* (The Analysis of Position) and gave the first comprehensive description of the fundamental concepts of combinatorial topology.[7] It became a standard reference work for many years and constituted Veblen's most important contribution to graph theory.

In his preface, Veblen recorded his indebtedness to Philip Franklin (see Chapter 3), who "assisted with . . . the manuscript, the drawings, and the proof-sheets". The following paragraph, which opens the preface, succinctly set the scene for the book's content and indicated the modest and unassuming personality of its author:

> The Cambridge Colloquium lectures on Analysis Situs were intended as an introduction to the problem of discovering the *n*-dimensional manifolds and characterizing them by means of invariants. For the present publication the material of the lectures has been thoroughly revised and is presented in a more formal way. It thus constitutes something like a systematic treatise on the elements of Analysis Situs. The author does not, however, imagine that

it is in any sense a definitive treatment. For the subject is still in such a state that the best welcome which can be offered to any comprehensive treatment is to wish it a speedy obsolescence.

The work was divided into five chapters, the first one being "Linear graphs". In it, Veblen established his fundamental definitions by building on the following basic parameters for linear graphs and using the complete graph K_4 as an example; this was the same graph as in his earlier paper, but with different terminology:

a 0-*dimensional simplex* is a single point,
a 1-*dimensional simplex* is a segment or edge,
a 0-*dimensional cell* (0-cell) is an end or vertex,
a 1-*dimensional cell* (1-cell) consists of the points of a segment,
a 0-*dimensional complex* is the set of distinct 0-cells,
a 1-*dimensional complex* is a linear graph.

Veblen then developed some 19th-century algebraic ideas put forth by G. R. Kirchhoff (who had written on electrical networks) and J. B. Listing (who had written on topological complexes), as refined by Henri Poincaré. In order to calculate the currents in a network, Kirchhoff had introduced the idea of a *spanning tree* (a tree that includes every vertex of the network) and from it obtained a "fundamental set of cycles" that enabled him to find all the currents.[8] Citing Kirchhoff's work, Veblen showed that the number of cycles in a fundamental set is related to the rank of the corresponding incidence matrix.

More generally, Veblen obtained incidence matrices for these multidimensional structures and discussed the theory of the n-cell, regular complexes, manifolds, and the associated dual complexes. He also included descriptions of the *rank* $V-1$ and the *nullity* $E-V+F$ of a map or planar graph, where V, E, and F are the numbers of vertices, edges, and faces (regions). These concepts of rank and nullity would later play a central role in the work of Hassler Whitney (see Chapter 4).

Following his successful series of lectures, Veblen turned his attention to other areas of mathematics. After the publication in 1915 of Einstein's general theory of relativity, he became interested in differential geometry, and from 1922 most of his publications were on this subject and on its connections with relativity. These investigations led to important applications, and atomic physicists were later to make use of his discoveries.

World War I interrupted Veblen's career, as we shall see, and on his return to Princeton he quickly became respected as a leading geometer. Because of his achievements, many graduate students applied to study there or to be employed by the mathematics faculty. One of these students was Philip Franklin and, as his postgraduate supervisor, Veblen was undoubtedly instrumental in Franklin's choice of thesis subject and in his continuing interest in graph theory and the four color problem (see Chapter 3).

Apart from his published work, Veblen's contributions to graph theory were through the direct influence that he had on other mathematicians—his near-contemporary, George Birkhoff, in particular. Their academic lives were closely linked through their mathematical work, and through their mutual interest in the development of American scholarship and of its standing in the world.

GEORGE D. BIRKHOFF

George David Birkhoff was the leading American mathematician of his time.[9] Of Dutch extraction, he was born in 1884 in Overisel, Michigan. From the age of 9, he showed a considerable aptitude for mathematics, and his solution to a problem in number theory appeared in the *American Mathematical Monthly* when he was just 15 years old.

In 1902, Birkhoff attended the University of Chicago where he met Oswald Veblen, then a graduate student. Despite their very different personalities and their disagreements on various issues, they became and remained close friends. In time, Veblen was to influence Birkhoff's early work and selection of research subject matter, including graph theory and the four color problem.

After just a year in Chicago, Birkhoff transferred to Harvard University, being awarded a bachelor's degree in 1905 and a master's degree in 1906. Returning to the University of Chicago in 1905 for postgraduate study under E. H. Moore, he earned his doctoral degree in 1907 for a thesis on the properties and applications of certain ordinary differential equations, work that was heavily influenced by the mathematics of Henri Poincaré.

Following the award of his doctorate, Birkhoff took up an appointment in 1907 as an instructor of mathematics at the University of Wisconsin in Madison. In 1909, he moved to Princeton University, and two years later he was appointed professor there and conducted research

George D. Birkhoff (1884–1944).

mainly in dynamics and mathematical physics. In 1912, he returned to Harvard as an assistant professor, was promoted to full professor in 1919, and remained there for the rest of his life.[10] In addition to his work on map coloring, Birkhoff made substantial contributions to many other mathematical areas, including dynamical systems, ergodic theory, differential and difference equations, the calculus of variations, and the three-body problem.

Chromatic Polynomials

At the same time as Veblen's paper on modular equations appeared, his colleague George Birkhoff was publishing his first major contribution to the coloring of maps. Birkhoff's interest and subsequent mild obsession with the four color problem had been triggered by attending Veblen's seminar in *Analysis Situs* when they were together at Princeton. His son Garrett, who also became a prominent mathematician, later recalled that:[11]

> The four color problem was one of my father's hobbies. I remember that all through the 1920's, my mother was drawing maps that he would then proceed to color. He was always trying to prove the four color theorem.

Dismissing as probably apocryphal a story that his mother had been requested by her new husband to prepare maps for him to color while on honeymoon, he continued:

> A related, definitely true story concerns [Solomon] Lefschetz. He came to Harvard—this must have been around 1942—to give a colloquium talk. After the talk my father asked Lefschetz, "What's new down at Princeton?" Lefschetz gave him a mischievous smile and replied, "Well one of our visitors solved the four color problem the other day." My father said: "I doubt it, but if it's true I'll go on my hands and knees from the railroad station to Fine Hall." He never had to do this.

In later life, the older Birkhoff was to rue the time and effort that he had devoted to the four color problem, even though it had always been a keen ambition of his to find a solution. Many years later, Hassler Whitney (see Chapter 4) recalled:[12]

> In the early 1930s, when I was at Harvard, exploring the problem among other things, Birkhoff told me that every great mathematician had studied the problem, and thought at some time that he had proved the theorem. In this period I was often asked when I thought the problem would be solved. My normal response became "not in the next half century."

Birkhoff's first paper on the subject was "A determinant formula for the number of ways of coloring a map".[13] Here he suggested a new avenue of investigation by taking a quantitative approach. This led to the introduction of the *chromatic polynomial*, which counts the possible colorings of the regions of a map when a given number of colors are available.

George D. Birkhoff: *A determinant formula for the number of ways of coloring a map* (1912)

In how many ways can a given map be colored with λ colors? Birkhoff's paper included two examples.

The first map has three mutually adjacent regions (ignoring the outside region): here, region A can be colored with any of the λ colors, region B can then receive any of the remaining $\lambda-1$ colors, and region C can have any of the remaining $\lambda-2$ colors. The total number of ways of coloring the map is therefore

$$P(\lambda)=\lambda(\lambda-1)(\lambda-2)=\lambda^3-3\lambda^2+2\lambda.$$

So if five colors are available, then the number of colorings is $P(5)=60$.

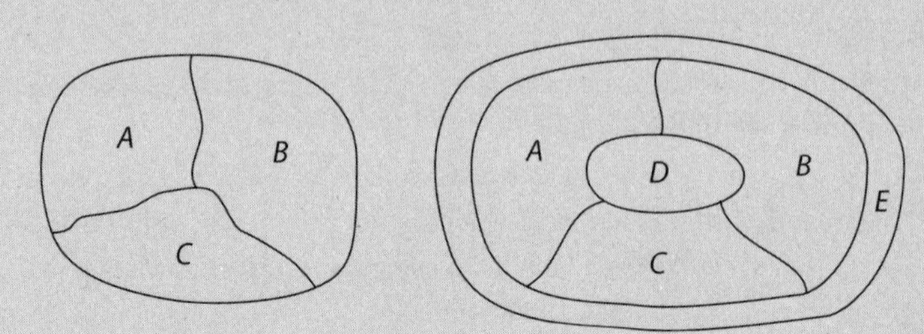

Birkhoff's second map has five regions: here, region A can be colored with any of the λ colors, B with $\lambda-1$ colors, C with $\lambda-2$ colors, and D and E with $\lambda-3$ colors each, so

$$P(\lambda)=\lambda(\lambda-1)(\lambda-2)(\lambda-3)^2=\lambda^5-9\lambda^4+29\lambda^3-39\lambda^2+18\lambda.$$

So if five colors are available, then the number of colorings is $P(5)=240$.

In every case, as Birkhoff discovered,

> The number of ways of coloring the given map M in λ colors ($\lambda=1, 2, \ldots$) is given by a polynomial $P(\lambda)$ of degree n, where n is the number of regions of the map M.

This polynomial is now known as the *chromatic polynomial* of the map.

Birkhoff also obtained the following formula for its coefficients:

$$P(\lambda)=\sum_i \lambda^i \sum_k (-1)^k\,(i, k).$$

The two summations here extend from $i=1$ to n, and from $k=0$ to $n-i$, and (i, k) is

> the number of ways of breaking down the map M in n regions to a submap of i regions by k simple or multiple coalescences.

The boundary conditions given were

$$(i, k)=0 \text{ for } k>n-i, \quad (n, 0)=1, \quad \text{and} \quad (i, 0)=0 \text{ for } i<n.$$

However, Birkhoff warned that "the value of (i, k) is not immediately obtained". He indicated that more complicated maps would require considerable computation to determine $P(\lambda)$ from the above equation or by inspection of the map itself.

Birkhoff's main objective was to prove that $P(4) > 0$, but he was unable to achieve this. Although he managed to find some general properties of $P(\lambda)$, his methods, involving determinants, were cumbersome. Twenty years later, Hassler Whitney discovered a simpler procedure for determining the coefficients of chromatic polynomials (see Chapter 4).

Birkhoff continued to work on chromatic polynomials for the rest of his life. In 1930, he wrote a paper[14] in which he proved that, if λ is any positive integer other than 4, then, for a map with n (≥ 3) regions,

$$P(\lambda) \geq \lambda(\lambda - 1)(\lambda - 2)(\lambda - 3)^{n-3}.$$

This is a tantalizing result, because if it could also be proved for $\lambda = 4$, then the four color problem would be solved!

In 1934, Birkhoff followed this with a paper[15] in which he considered the roots of the equation $P(\lambda) = 0$, in order to place the problem in a wider context. Finally, in the early 1940s, he teamed up with Daniel C. Lewis to write a major work on chromatic polynomials (see Chapter 5).

Reducibility

In 1913, Birkhoff proved Henri Poincaré's "last geometric theorem", a restricted form of the three-body problem of finding the simultaneous motion of the sun, earth, and moon, which Poincaré had been unable to solve. Few scholars knew and understood Poincaré's work better than Birkhoff, whose efforts in this area helped to establish his international fame. He soon came to be regarded as North America's leading mathematician, and his proof remains widely cited today.

Also published in 1913 was Birkhoff's pioneering paper, "The reducibility of maps",[16] which would prove significant in the development of the solution to the four color problem. In it he took a more systematic approach to map coloring, by first defining a *configuration* in a map to be a collection of regions that are surrounded by an outside ring of regions.

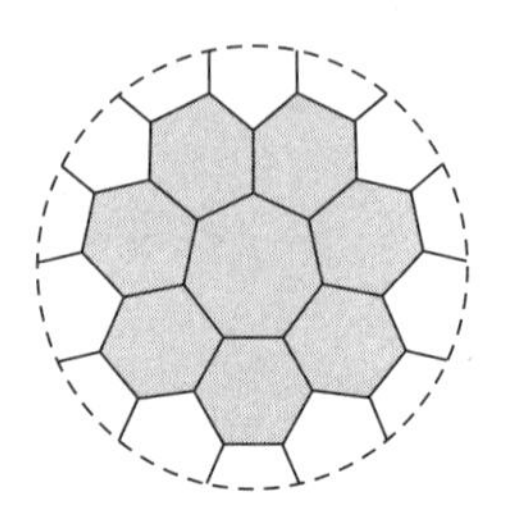

A configuration of regions in a map.

Birkhoff then investigated which colorings of the regions in the surrounding ring can be extended to the regions inside it—either directly or by interchanging pairs of colors, as described by Kempe (see Chapter 1). When this can be done, the configuration is said to be *reducible*; otherwise, it is *irreducible*. It follows that a reducible configuration is an arrangement of regions that cannot occur in a minimal counter-example to the four color theorem—that is, in a map that requires more than four colors, but for which all maps with fewer regions can be colored with four colors.

Kempe had shown that digons, triangles, and quadrilaterals are all reducible. Birkhoff now carried this idea much further by systematically investigating for reducibility those configurations that are surrounded by rings with up to six regions.

George D. Birkhoff: *The reducibility of maps* (1913)

This pioneering paper was published in the *American Journal of Mathematics*. In his introduction, Birkhoff maintained that all previous work on the four color problem could be stated in terms of the following four "reductions" or simplifications:

If more than three boundary lines meet at any vertex of a map, the coloring of the map may be reduced to the coloring of a map of fewer regions.

If any region of a map is multiply-connected, the coloring of the map may be reduced to the coloring of maps of fewer regions.

If two or three regions of a map form a multiply-connected region, the coloring of the map may be reduced to the coloring of maps of fewer regions.

If the map contains any 1-, 2-, 3- or 4-sided region, the coloring of the map may be reduced to the coloring of a map of fewer regions.

Birkhoff concluded his introductory remarks by declaring that his purpose was

> to show that there exist a number of further reductions which may be effected with the aid of the notion of *chains* due to Kempe.

To this end, he developed Kempe's arguments by systematically studying rings of regions in his search for reducible configurations.

Birkhoff considered a ring R of regions that divides a map M into two sets of regions, M_1 and M_2, so that $M = M_1 + M_2 + R$. The colorings of the maps $M_1 + R$ and $M_2 + R$ can then be combined to give a complete coloring of M, provided that the two separate colorings can be made to agree on the ring R. By applying the method of Kempe chains to colorings of $M_1 + R$ and $M_2 + R$, Birkhoff was able to prove that no irreducible map can contain a ring with four regions.

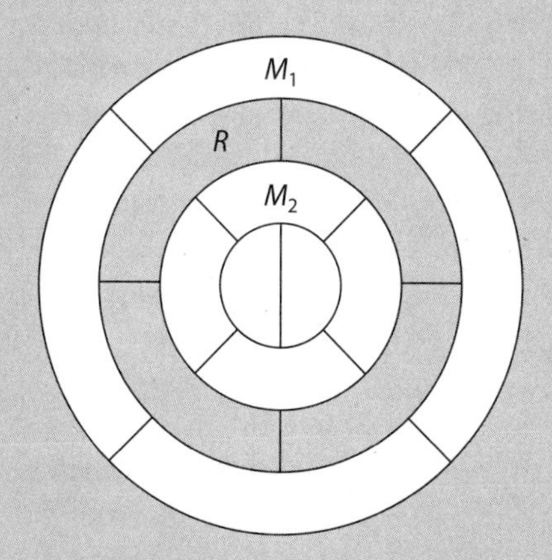

He next repeated the procedure for rings with five regions and showed that for such rings the configuration is always reducible, except when the ring surrounds a single pentagon. He had less success with rings with six regions, but managed to prove that a particular arrangement of four pentagons is reducible—it is now known as the *Birkhoff diamond*. Here, as we shall see in Chapter 6, there are thirty-one essentially different ways of coloring the outside ring,

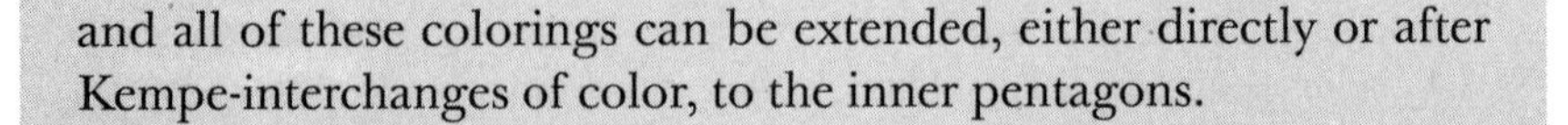
and all of these colorings can be extended, either directly or after Kempe-interchanges of color, to the inner pentagons.

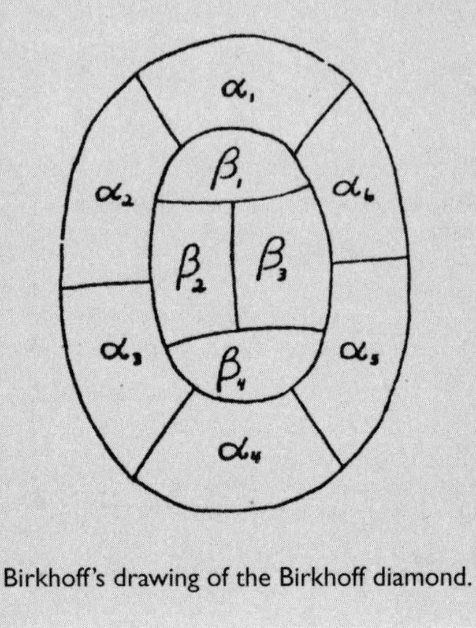

Birkhoff's drawing of the Birkhoff diamond.

Included in Birkhoff's paper was a list of possible outcomes of the four color problem, summarized as follows:

1. There are maps that cannot be colored in four colors.
2. All maps can be colored in four colors, and a list of reducible rings can be found, at least one of which appears in every map.
3. All maps can be colored in four colors, but only by means of more complicated reductions.

His desire for a listing of possible reducible configurations provided the basis upon which future investigators would continue his work. Several mathematicians would use Birkhoff's results on rings with five regions, one being Philip Franklin (see Chapter 3). A full analysis of the rings with six regions was not completed until the 1940s, as we shall see in Chapter 5.

George Birkhoff was a complex man. As a widely respected figure in his homeland and abroad, he received many honors and awards from around the world. He was also warmly praised for the encouragement and support that he gave his research students and colleagues, but he seems not to have favored women in academia. In the 1930s, he was criticized for anti-Semitism, as we shall see.

WORLD WAR I

As a British Dominion, Canada declared war on Germany on August 4, 1914, with the United States following on April 6, 1917, after more than two years of action in Europe. In doing so, America then initiated a deployment program of academics from universities and colleges across the country to assist in the war effort. Included in this effort were nearly 200 mathematicians who became engaged in war service.

Of those mathematicians who were called to arms, the majority were employed in research into war-related technologies, while others saw active duty with the army, navy, or air force, or acted as instructors and technical administrators. Whereas most were stationed in North America, some served in France, Italy, and Britain. Included among these in 1917 were two who feature in our story:

> Oswald Veblen (aged 37), professor of mathematics at Princeton University, was commissioned as a major in the US Army and was assigned to research in ordnance at the Aberdeen Proving Ground in Maryland.
>
> Philip Franklin (aged 19), an undergraduate at the College of the City of New York, worked alongside Veblen in the range-finding section at the Aberdeen Proving Ground.

We have found no mention of George Birkhoff among the American mathematicians who assisted in the war effort—in particular, his name does not appear in the list of *American Mathematicians in War Service*, later published by the American Mathematical Society.[17] Nor is there any mention of Paul Wernicke, even though he had become a naturalized American citizen some years earlier. While continuing to teach in Kentucky, he had been in charge of his college's military training, holding a commission as a colonel in the Kentucky militia.

This involvement of academics in the practical world of war-related technologies resulted in many pure mathematicians becoming exposed to more applied pursuits. Once the war was over, life then returned to normal for these mathematicians, with most resuming their academic positions, while others, like Franklin, were able to complete their degrees.

* * * * *

In the quest for a solution to the four color problem, the contributions of Paul Wernicke (on unavoidable sets) and George Birkhoff (on reducible configurations) proved to be of fundamental importance, while the more abstract approach of Oswald Veblen helped to position the study of graphs in a more mathematical context. In the 1920s, valuable advances were made by Franklin and others, as we shall discover in Chapter 3, but it would not be until the 1930s that the American graph theory scene became transformed forever.

Chapter 3
The 1920s

The 1920s was a decade of increasing involvement in graph theory in America, and in coloring problems in particular, as American mathematicians began to follow the directions set by Oswald Veblen and George Birkhoff. One of these was Philip Franklin, a postgraduate student of Veblen's whose doctoral thesis was on map coloring. Others that we discuss here are H. Roy Brahana, the "forgotten mathematician" Howard Redfield, and Clarence Reynolds Jr. We also leap forward to look briefly at further contributions to map coloring in the late 1930s.

PHILIP FRANKLIN

Philip Franklin was a quiet and unassuming man who seemed less able or willing than others to promote himself, but who nevertheless provided undemonstrative encouragement and support to his students. He was also loyal to his colleagues, and to the departments to which he was

Philip Franklin (1898–1965).

appointed, but because he preferred to maintain his privacy, there is less biographical information available on him than on other mathematicians we have featured.

In 1914, Franklin enrolled at the City College of New York and graduated with a bachelor's degree in 1918. He then undertook postgraduate study at Princeton University, where he received a master's degree in 1920 and his doctorate in 1921. His supervisor was Oswald Veblen, and his chosen research topic was the four color problem.

Franklin remained for a further year at Princeton as an instructor in mathematics, before spending two years at Harvard University as Benjamin Peirce Instructor. In 1924, he transferred to the Massachusetts Institute of Technology, becoming an assistant professor there in 1925 and an associate professor five years later. He was promoted to full professor in 1937, a position that he held until his retirement in 1964. A highly praised teacher, he wrote eight undergraduate textbooks on a range of topics.

As we have seen, World War I interrupted Franklin's undergraduate studies when he joined the army's ordnance Aberdeen Proving Ground in Maryland. While there, he met another member of Veblen's group, Norbert Wiener, later the inventor of cybernetics. A fellow mathematician was David Widder, who had a bunk in the same barracks as Franklin and Wiener and who later recalled:[1]

> I learned a lot from these enthusiasts, but at times they inhibited sleep when they talked mathematics far into the night. On one occasion I hid the light bulb, hoping to induce earlier quiet.

A child prodigy, Wiener had enrolled in college at the age of 11 and obtained his doctoral degree from Harvard when he was only 18. But he failed to get an appointment there, suspecting that Harvard's newly appointed assistant professor, George Birkhoff, was partly responsible for this. Birkhoff was one of a small number of Harvard mathematicians in the 1920s who warned Wiener against pursuing the then-topical subject of potential theory, so as to leave the field clear for other Harvard scholars, but Wiener also claimed that Birkhoff showed him special antipathy as a Jew. Wiener eventually found a position in the mathematics faculty at MIT, which he took up in 1919, and remained there for forty-five years until his death in 1964.

Philip Franklin would later become Wiener's brother-in-law, when he married Wiener's sister Constance, who was a mathematician in her own right. Both Franklin and Wiener were involved in introducing topology to MIT, as later recalled by their colleague Dirk Struik:[2]

Since Franklin brought topology to MIT in his "analysis situs" form, and Wiener in its "point-set-Lebesgue" form, we see that it came to the Institute through two brothers-in-law.

The Four Color Problem

After the war, Franklin returned to academic life and was working on his doctorate, and on November 17, 1920, he presented some of his results on the four color problem to the National Academy of Sciences. Soon after this, following the submission of his doctoral thesis in 1921, he published a major paper on the topic in which he developed the work of Wernicke on unavoidable sets and that of Birkhoff on reducible configurations.[3] This paper made a significant contribution to the ideas of unavoidable sets and reducibility as avenues toward solving the four color problem. It was the first in a continuing line of papers that adopted this approach over the years, as we shall see. Franklin's development of the subject was a model of clarity and precision.

Philip Franklin: *The four color problem* (1922)

The following result is similar to the one used by Kempe (see Chapter 1) and is a consequence of Euler's formula:

Counting theorem: If, for each k, C_k is the number of k-sided regions in a cubic map, then

$$4C_2 + 3C_3 + 2C_4 + C_5 - C_7 - 2C_8 - 3C_9 - 4C_{10} - \cdots = 12.$$

So every cubic map that contains no digon, triangle, or square ($C_2 = C_3 = C_4 = 0$) must contain at least twelve pentagons, and it follows that all maps with up to 12 regions can be colored with four colors. As we saw in Chapter 2, Wernicke had further proved that such a map must contain a pentagon adjacent to another pentagon, or a pentagon adjacent to a hexagon, and Franklin now showed that it must contain at least one of the following:

a pentagon adjacent to two other pentagons,
a pentagon adjacent to a pentagon and a hexagon,
a pentagon adjacent to two hexagons.

This result gave rise to a new unavoidable set with nine configurations.

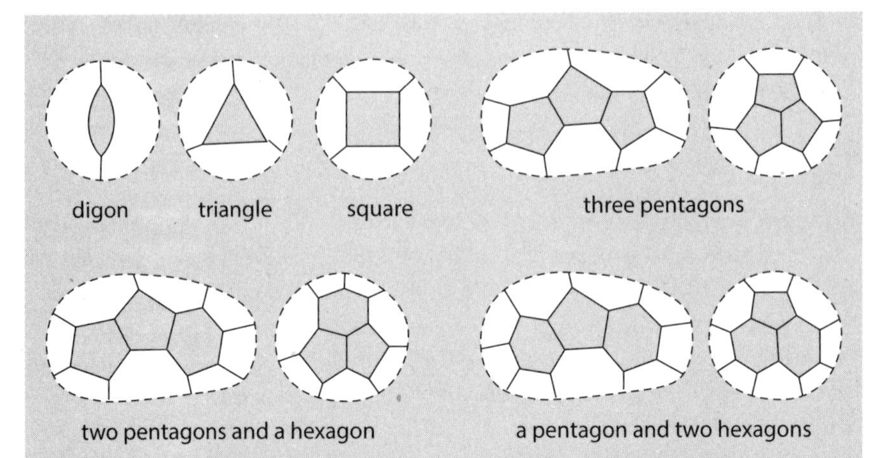

Franklin's unavoidable set.

Franklin also proved that no hexagon can have three consecutive pentagons as neighbors, and then introduced some new reducible configurations, including

an n-sided polygon in contact with $n-1$ pentagons,
a pentagon in contact with three pentagons and one hexagon,
a pentagon in contact with two pentagons and three hexagons,
a hexagon in contact with four pentagons and two hexagons.

We have just seen that all maps with up to 12 regions can be colored with four colors. By a detailed analysis of maps that contained no known reducible configurations, such as those of Birkhoff and the four listed above, Franklin was able to improve on this result:

> Every map on the plane or sphere containing 25 or fewer regions can be colored in four colors.

Defining a map to be *irreducible* if it is a minimal counter-example to the four color theorem, Franklin deduced that an irreducible map must have at least 26 regions, and continued:

> The question naturally arises whether 25 is the greatest number for which we can prove such a theorem as the above on the basis of the reductions already described. While an exact answer to this question is lacking, it is evident that the smallest number of regions in a map not containing any of these known reducible configurations is not con-

siderably above 25, as we can construct a map with a small number of regions not containing any of them.

Franklin's example had 42 regions and can be colored with four colors.

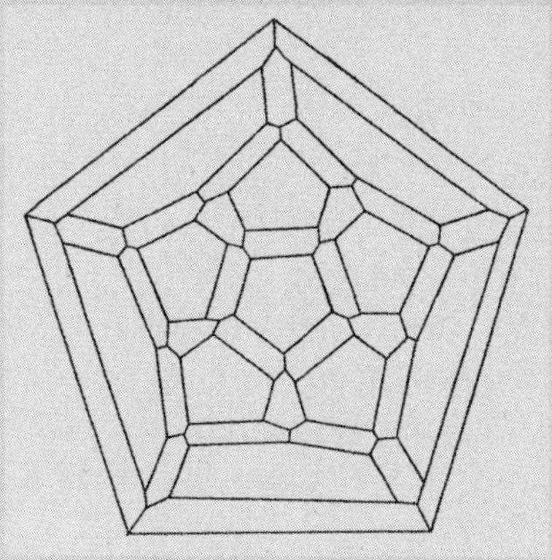

Philip Franklin's 1922 paper was an important step along the road that other map colorers were to follow. But his second publication on graph theory, "The electric currents in a network", had nothing to do with map coloring.[4]

Electrical Networks

In our discussion of Veblen's Colloquium Lectures on Analysis Situs (see Chapter 2), we outlined Kirchhoff's methods for determining the currents in an electrical network by obtaining a "fundamental set of cycles" from which they could conveniently be calculated. Franklin examined this approach and, after commenting that Kirchhoff's paper had been "essentially the first contribution to the study of analysis situs of the linear graph", he discovered an alternative proof of Kirchhoff's result that was "somewhat shorter than his and also brings to light the mathematical nature of the result".

Franklin presented his proof to the American Mathematical Society on October 25, 1924, crediting Veblen's lectures for the terminology and treatment of graphs, and adopting his example of the complete graph K_4. Further work along these lines was subsequently carried out by Ronald M. Foster, a mathematician at the Bell Telephone Laboratories.[5]

A Six Color Problem

A few years later, in 1934, Franklin returned to map coloring when he confirmed a possible exception to the Heawood conjecture for non-orientable surfaces. As we saw in Interlude A, Heinrich Tietze used a version of Euler's formula to prove that, when $q \geq 1$, the regions of every cubic map drawn on the non-orientable surface $\mathcal{N}_q$ can be colored with

$$\mathcal{T}(q) = \left\lfloor \tfrac{1}{2}\left(7 + \sqrt{1 + 24q}\right) \right\rfloor$$

colors; for example, the regions of every map on a Klein bottle (where $q = 2$) can be colored with seven colors.

Heawood's conjecture for non-orientable surfaces asserted that there are maps on $\mathcal{N}_q$ that require this number of colors. But although Tietze had found a map on the Klein bottle requiring six colors, he was unable to find one that needs seven. Franklin proved definitively that there are no such maps, so that the Heawood conjecture fails when $q = 2$. He also presented the following map that requires six colors.

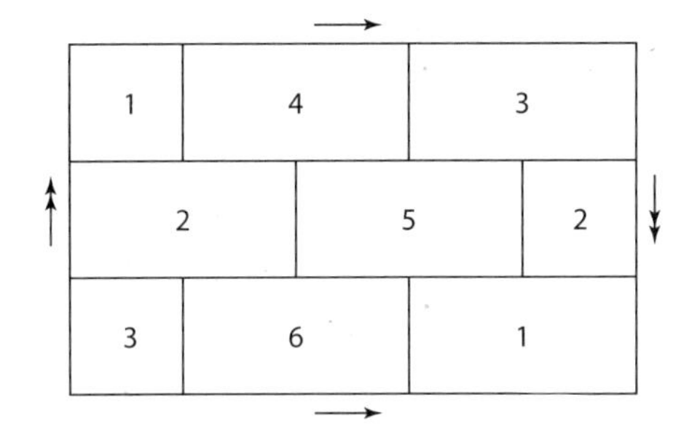

Franklin further remarked that because "Heawood's formula is incorrect for this surface", it "may also fail in other cases". But in the succeeding years, Isidore Kagno of Columbia University proved that Tietze's formula gives the correct values of 7, 8, and 9, when $q = 3$, 4, and 6 (see Chapter 5), and Donald Coxeter and R. C. Bose confirmed the corresponding results for $q = 5$ and 7.[6] Then, in 1952, Gerhard Ringel proved that Franklin had indeed found the only instance where Tietze's formula fails (see Chapter 6).

In Interlude A, we also met Tietze's version of Heawood's conjecture which asserts that the simplest non-orientable surface on which the complete graph K_n can be drawn without crossings is $\mathcal{N}_q$, where

$$q = \left\lceil \tfrac{1}{6}(n-3)(n-4) \right\rceil.$$

Again, this result is true, except in a single case: when $n=7$, this formula gives $q=2$, whereas the correct value is 3.

Franklin's paper "A six color problem" was published in the 1934 edition of the *Journal of Mathematics and Physics* and was presented to the American Mathematical Society that same year.[7] In subsequent years, he wrote two further papers on map coloring, which we look at later in this chapter, but first we turn our attention to another American mathematician who colored maps in the 1920s.

H. ROY BRAHANA

Henry Roy Brahana attended Dartmouth College in New Hampshire and Princeton University, where he obtained his doctorate in 1920 on *Systems of Circuits on Two-Dimensional Manifolds*, under the supervision of Oswald Veblen. He then transferred to the University of Illinois at Urbana–Champaign, where he became a professor and supervised twenty-seven doctoral students. He retired in 1963.

Brahana's contributions to graph theory amounted to three papers. The first of these, "A proof of Petersen's theorem",[8] was published at a time when American scholars were beginning to take an interest in graph theory and coloring problems. Written in 1917 during his postgraduate

H. Roy Brahana (1895–1972).

studies at Princeton, it provided a simpler proof of Petersen's factorization theorem (see Interlude A).

His second paper, "The four color problem",[9] which appeared in 1923, outlined the problem's history from Francis Guthrie's initial query in 1852 to the early 1920s. In his survey, Brahana cited most of the mathematicians who had worked on coloring problems, including De Morgan, Cayley, Kempe, Tait, Heawood, Wernicke, Petersen, Veblen, Birkhoff, Errera, and Franklin. His paper included some pertinent observations, such as:

> The problem is still unsolved. It has afforded many mathematicians experience and very little else.

It was also in 1923 that Brahana presented an important result on the topology of surfaces (see Interlude A):[10]

> *Brahana's theorem:* Every surface is topologically equivalent to either some orientable surface S_g or some non-orientable surface N_q.

This implies that any surface obtained from a sphere by adding handles and cross-caps can be produced by adding handles only or cross-caps only.

Brahana's third paper, "Regular maps on an anchor ring",[11] was published in 1926. On a sphere, the "regular maps"—those whose regions are all the same—are obtained by projecting the five regular polyhedra onto it. Brahana was interested in the corresponding maps on an "anchor ring" or torus and, as he observed:

> any regular map on an anchor ring [has] triangular, quadrangular or hexagonal regions. In none of these cases is there a restriction on the number of regions. If the regions are triangular they appear six at a vertex; if quadrangular, four at a vertex; and if hexagonal, three at a vertex. Since two adjacent vertices are joined by a line and two adjacent regions are separated by a line there is a sort of duality between the maps of triangles and the maps of hexagons. The quadrangular maps are self-dual.

He consequently restricted his investigations to maps with quadrangular and hexagonal regions only, seeking conditions for their existence. Invoking the algebraic theory of abstract groups, he obtained the following results:

> A regular map with n quadrangles exists if and only if there is a group of order $4n$ that is generated by two elements of orders 2 and 4 whose product has order 4.

> A regular map with n hexagons exists if and only if there is a group of order $6n$ that is generated by two elements of orders 2 and 3 whose product has order 6.

In 1901, the American group-theorist George A. Miller had found the values of n that produce regular hexagonal maps.[12] Brahana obtained the corresponding values for regular quadrangular maps, and proved that each corresponding group yields a unique map. Further work on this topic was subsequently carried out by Richard P. Baker, an Englishman who taught mathematics for many years at the University of Iowa.[13]

J. HOWARD REDFIELD

The name of John Howard Redfield was largely unknown until the 1960s, and yet his results anticipated one of the most important achievements in combinatorial mathematics in the mid-20th century—the enumerative work of George Pólya in the 1930s. It is only more recently that Redfield's contributions to the subject have been recognized.[14]

Howard Redfield had a most unusual student career. After gaining a bachelor of science degree from Haverford College near Philadelphia, he then took a second bachelor's degree in civil engineering from MIT before completing a doctorate in romance languages from Harvard University. His subsequent employment as a civil engineer led him to an interest in the mathematical theory of elasticity, and in

J. Howard Redfield (1879–1944).

turn to other areas of mathematics such as mathematical logic and the theory of knots. Particularly inspired by topics in Percy A. MacMahon's two-volume treatise on *Combinatory Analysis*,[15] he began to investigate counting problems that involved symmetry. Such considerations led him to explore the connections between enumeration and algebraic groups of symmetries, and in 1927 he wrote the following remarkable paper on the subject, which included applications to the theory of graphs.[16]

J. Howard Redfield: *The theory of group-reduced distributions* (1927)

An example of the type of problem that interested Redfield is the following:

> Required the number of distinct configurations which can be obtained by placing a solid node • at each of four vertices of a cube, and a hollow node ○ at each of the four remaining vertices, configurations differing only in orientation not being regarded as distinct.

Without this last restriction there would be $\binom{8}{4} = 70$ different ways of selecting the four solid nodes, but if we regard two placings of the vertices as the same when we can rotate the cube to get from one to the other, then this number is reduced to 7, as we shall see.

Redfield first set the scene:

> In view of the similarity which will be admitted to hold between the subject matters of the Theory of Finite Groups and of Combinatory Analysis, it is somewhat surprising to find that in their literatures the two branches have proceeded on their separate ways without developing their interrelationship, and with scarcely any reference to one another beyond the use by each of certain very elementary results of the other.

To solve such problems, Redfield introduced a polynomial expression which he called the *group reduction function* of a finite group G of operations acting on a set of elements S. This polynomial records the lengths of cycles associated with each element of G, regarded as a permutation of the members of S. For this problem, S is the set of eight nodes and the group G consists of the 24 rotations of the cube, which can be classified as follows:

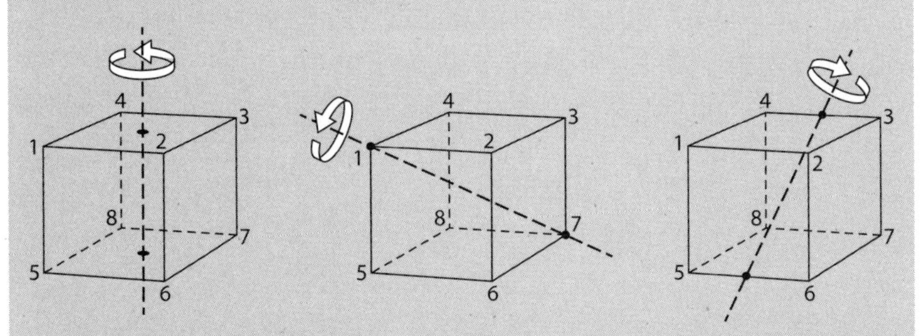

one identity element that fixes all eight nodes, giving eight 1-cycles, (1)(2)(3)(4)(5)(6)(7)(8): he recorded this as s_1^8,

three rotations through 180° about opposite faces, each one interchanging four pairs of nodes (2-cycles) such as (13)(24)(57)(68): he recorded these as $3s_2^4$,

six rotations through ±90° about opposite faces, each one with two 4-cycles such as (1234)(5678): he recorded these as $6s_4^2$,

eight rotations through ±120° about a line through a pair of opposite vertices, each one fixing two nodes (1-cycles) and with two 3-cycles such as (1)(7)(245)(368): he recorded these as $8s_1^2s_3^2$,

six rotations through 180° about a line joining the midpoints of opposite edges, each one interchanging four pairs of nodes such as (17)(28)(34)(56): he recorded these as $6s_2^2$.

The group reduction formula is the average of these, which is

$$\tfrac{1}{24}\left(s_1^8 + 9s_2^4 + 8s_1^2s_3^2 + 6s_4^2\right).$$

Redfield then replaced each symbol s_k by $x^k + y^k$, where x corresponds to a solid node and y corresponds to a hollow one, giving

$$\tfrac{1}{24}((x+y)^8 + 9(x^2+y^2)^4 + 8(x+y)^2(x^3+y^3)^2 + 6(x^4+y^4)^2)$$
$$= 1x^8 + 1x^7y + 3x^6y^2 + 3x^5y^3 + 7x^4y^4 + 3x^3y^5 + 3x^2y^6 + 1xy^7 + 1y^8.$$

Here, the coefficient of x^ty^{8-t} is the number of placings with t solid nodes and $8-t$ hollow ones, and so the answer to Redfield's original problem is 7. Notice also that the sum of the coefficients is 23, which is the total number of different placings of the nodes, taking into account the rotations of the cube.

Much of Redfield's paper was difficult to read and involved complicated notation and terminology—for example, in his enumera-

REDFIELD: *The Theory of Group-Reduced Distributions.* 443

The actual configurations, shown below, cannot be determined by the methods of the present theory, but must be found, as in all other cases, by detailed consideration of the groups involved, and this may of course be very laborious, except in simple cases, or where special devices are available.

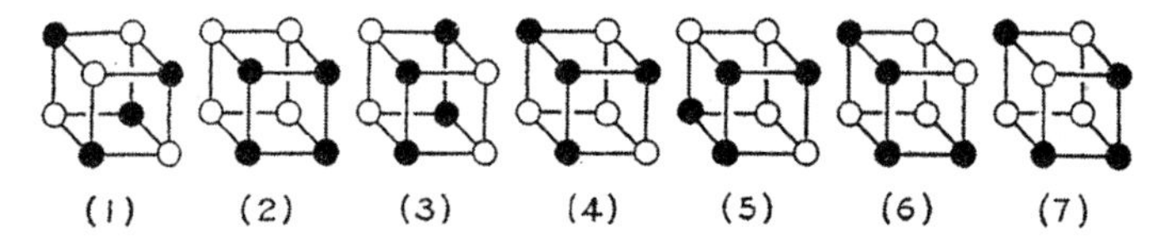

In connection with the present example we may note without proof certain other simple results obtainable.

Thus if in V we substitute $x^r + y^r$ for every s_r, we obtain the polynomial

$$x^8 + x^7y + 3x^6y^2 + 3x^5y^3 + 7x^4y^4 + 3x^3y^5 + 3x^2y^6 + xy^7 + y^8,$$

in which the coefficient of $x^t y^{8-t}$ enumerates the distinct configurations possible with t nodes • and $8 - t$ nodes ∘

The sum of the coefficients in the above expression is 23, which is the total number of configurations when the numbers of nodes of the two colors are not specified. This enumeration is also effected by substituting 2 for every s_r in V. Similarly if k colors are available we substitute k for every s_r; thus with 3 colors there are $(1/24)(3^8 + 9.3^4 + 8.3^4 + 6.3^2) = 333$ possible configurations.

If in V we put $1/(1 - x^r)$ for every s_r, we obtain the infinite series

$$1 + x + 4x^2 + 7x^3 + 21x^4 + 37x^5 + \cdots,$$

in which the coefficient of x^t enumerates the distinct configurations obtained by placing a zero or a positive integer at every vertex of the cube, subject to the condition that the sum of the 8 numbers is always t. For $t = 2$, the 4 configurations are

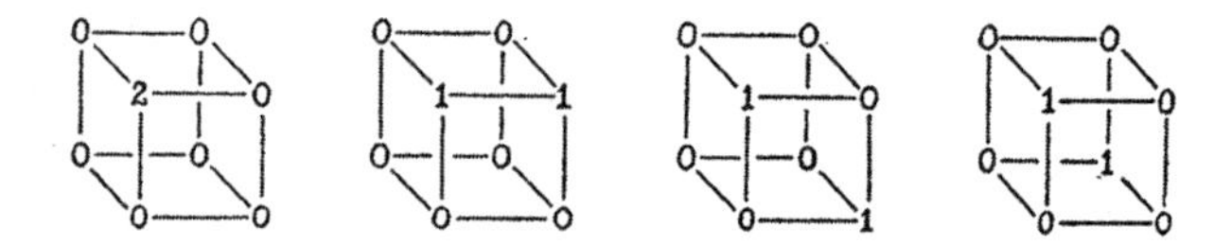

If in V we put 2 for every s_{2k} and 0 for every s_{2k+1}, we enumerate the configurations in which it is possible to change the color of every node into

A page from Redfield's paper.

tion of simple graphs with five vertices and five edges, he called them "symmetrical aliorelative dyadic relation-numbers". This may be why his paper lapsed into obscurity.

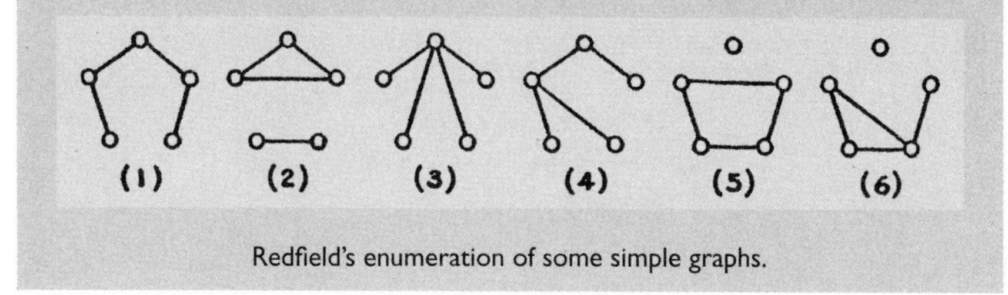

Redfield's enumeration of some simple graphs.

In the mid-1930s, unaware of Redfield's prior contributions to the subject, George Pólya independently wrote a number of papers on enumerative combinatorics. He was a Hungarian mathematician who taught from 1914 to 1940 in Zürich, but because of the political situation in Europe he migrated to the United States, where he taught for a short while at Brown University and Smith College, before he moved to Stanford University, where he spent the remainder of his long life. He is widely remembered for his writings on mathematics education, in particular his best-selling book, *How to Solve It*,[17] first published in German and then in English in 1945.

The most important of Pólya's contributions to the theory of enumeration was a groundbreaking paper in 1937 of over 100 pages,[18] written

George Pólya (1887–1985).

in German while he was still in Zürich. In it Pólya used the group reduction formula, here renamed the "cycle index", to enumerate various types of graphs and chemical molecules. Although some of his results had been anticipated by those of Redfield, he carried them much further, and this paper is of fundamental importance to the history of the subject. It was eventually translated into English in 1987.[19]

A TRIO OF MAP COLORERS

We have seen how a detailed analysis of maps that contained no known reducible configurations enabled Philip Franklin to prove that

> Every map on the plane or sphere containing 25 or fewer regions can be colored with four colors.

Over the next few years, three American mathematicians improved on this result—Clarence Reynolds Jr. raised the number of regions from 25 to 27 in 1927; then, a few years later, Philip Franklin (again) increased it to 31, and C. E. Winn raised it still further to 35. In order to do this, they sought "irreducible maps"—minimal counter-examples to the four color theorem. By discovering more and more information about such maps—for example, that they contained no reducible configuration then known—they hoped to be able either to construct them explicitly or to prove that they could not exist.

Clarence N. Reynolds Jr.

Clarence Newton Reynolds Jr. undertook his postgraduate studies at Harvard, under the direction of Maxime Bôcher and George Birkhoff, and was awarded his doctorate in 1919 for a dissertation on the solutions of differential equations. After leaving Harvard, he moved to the University of West Virginia, became head of the mathematics department in 1938, and remained there until his retirement in 1946.

After learning of the four color problem, possibly from Birkhoff, Reynolds published several papers on map coloring. The most important of these were two papers that he wrote around 1926, in which he developed the methods of earlier authors.[20] The first of these papers sets the scene:

> The problem of coloring in four colors the map of a simply connected closed surface has been reduced to the problem of coloring maps in which certain

Clarence N. Reynolds Jr.
(1890–1954).

> configurations, known as reducible configurations, are absent. In this paper we shall develop some methods of so analysing the known geometric reductions of our problem as to discover and to prove some of their more important implications.

Reynolds's aim was to investigate irreducible maps that excluded "the known geometric reductions" (reducible configurations) of Birkhoff and Franklin:

> Our fundamental method will be a systematic study of geometric operations which suffice to build any connected configuration of pentagons which exist in an irreducible map. Under these operations certain numerical topological characteristics are found to undergo well defined increments.

But his analysis was even more detailed than Franklin's had been, and he was also able to simplify the process by excluding a reducible configuration of the Belgian mathematician Alfred Errera (see Interlude B) that involved pairs of adjacent pentagons surrounded by hexagons. He continued:

> Linear relations between these increments imply homogeneous linear difference equations which yield certain homogeneous relations between our topological characteristics.

From these linear relations, Reynolds derived some inequalities between these topological parameters and then applied these inequalities to improve on Franklin's earlier result:

> Every map on the plane or sphere containing 27 or fewer regions can be colored in four colors.

This could also have been obtained without excluding Errera's configuration, but the passage was smoothed considerably by doing so. Reynolds also presented a map with 28 regions, which avoided the configurations of Birkhoff and Franklin, but not that of Errera, and observed that his result could not be improved further without excluding further reducible configurations.

In 1934, in his last paper on graph theory, "Circuits upon polyhedra",[21] Reynolds left the world of coloring to investigate when closed curves can be drawn on a convex polyhedron or map. In particular, he derived some necessary and sufficient conditions for the existence of closed curves that pass exactly once through every region, and of Hamiltonian cycles that pass along its boundaries, visiting every point exactly once.

Philip Franklin

Several years later, in 1938, Philip Franklin succeeded in improving Reynolds's result on map coloring, proving in his "Note on the four color problem"[22] that

> Every map on the plane or sphere containing 31 or fewer regions can be colored with four colors.

To do so, he investigated the properties of irreducible cubic maps with 32 regions or more. Observing that $3P=2B$, for a cubic map with B boundary lines and P meeting points, he substituted this into Euler's formula, $R+P=B+2$, where R is the number of regions, and deduced that $R=2+P/2$. It follows that P is even.

Franklin then showed that an irreducible map cannot contain any of the following new configurations:

> exactly six heptagons, and no region with more than seven boundaries,
>
> more than six heptagons, and no region with more than seven boundaries,
>
> one octagon, at least five heptagons, and no other region with more than seven boundaries,
>
> one region with more than eight boundary lines, or at most two regions with more than seven boundaries,
>
> at least three regions with more than seven boundaries.

For each of these, Franklin calculated that $P \geq 60$, and so, by the preceding formula, $R \geq 32$. It follows that any irreducible cubic map must have at least 32 regions, and therefore that any map with at most 31 regions can be colored with four colors, as he had claimed.

In his proof, Franklin was also able to involve some reducible configurations that C. E. Winn had just published. It is to Winn that we now turn our attention.

C. E. Winn

Charles Edgar Winn was another American mathematician who was captivated by coloring problems and who, between 1937 and 1940, published five papers on the subject. He spent most of his career at the Egyptian University in Cairo and, with his university colleague Ismail Ratib, contributed a paper[23] to the International Congress of Mathematicians in Oslo in 1936. In it they extended one of Errera's reductions of the four color problem. Errera had proved that any map containing no regions with more than six boundaries is reducible, and Ratib and Winn strengthened this by allowing the map to have at most one such region.

In 1937 and 1938, Winn wrote three papers on map coloring[24] in which he discovered several new reducible configurations, such as a heptagon that is adjacent to four consecutive pentagons. He also listed all the reducible configurations known up to that time. In 1939, he published a short note outlining the history of the four color problem.[25] But his most important contribution to the subject was the paper "On the minimum number of polygons in an irreducible map", published in 1940.[26] Here he found yet more reducible configurations, which—in conjunction with those of Birkhoff, Franklin, Errera, and Reynolds—enabled him to prove that an irreducible map must contain at least 36 regions, and hence that

> Every map on the plane or sphere containing 35 or fewer regions can be colored with four colors.

No further improvements along these lines would be made for another thirty years.

Philip Franklin

We conclude this chapter by returning once more to Philip Franklin. Around 1939, he lectured on the four color problem at the Galois Institute of Mathematics at Long Island University in New York. A revised version of his lecture soon appeared as a two-part expository paper[27] and provides an excellent summary of all the topics on map coloring that we have met so far, including:

a statement of the four color problem,

a history of the problem from its beginnings to the late 1930s,

Euler's formula for maps, and the consequent restriction of the problem to cubic maps with no digons, triangles, or quadrilaterals,

a proof of the five color theorem,

the definition of a reducible configuration (with several examples),

the fact that irreducible maps have at least 36 regions,

Kempe's conditions for maps to be colorable with two or three colors,

Tait's result on the coloring of the boundaries of a cubic map,

Heawood's results on the coloring of cubic maps,

Petersen's theorem on the existence of a 1-factor in a cubic graph,

coloring cubic maps with a Hamiltonian cycle,

coloring maps on orientable and non-orientable surfaces,

the connection with mutually neighboring regions,

the chromatic polynomial of a map.

Franklin concluded his impressive list of topics by citing a probabilistic argument of Heawood's which implied that if the number of regions exceeds 35, then the probability that an uncolorable map can exist is less than 1 in $10^{10,000}$—a powerful argument in favor of the truth of the four color theorem, but still a long way from a proof.

* * * * *

Whereas the mathematicians mentioned in this chapter made notable contributions to graph theory, none of them approached the significance of Hassler Whitney's achievements in the 1930s. Chapter 4 is devoted mainly to Whitney's deep and varied results, but first we pay a brief return visit to Europe.

Interlude B
Graph Theory in Europe 2

While American graph theorists in the early 20th century concentrated on map coloring, their European counterparts were mainly heading in other directions. In Hungary, Dénes König worked on factorization and matching, while writing popular books on mathematical recreations. In Belgium, the interests of Alfred Errera lay still with the four color problem, but in France, André Sainte-Laguë wrote an early introductory monograph on other aspects of graph theory, while Karl Menger of Austria presented an important "minimax theorem" on the connectivity of graphs, and Kasimierz Kuratowski of Poland proved a classic result on planar graphs.

With this Interlude, we begin to move away from the regions, boundaries, and meeting points of maps, toward the language of graphs, vertices, and edges.

DÉNES KÖNIG (HUNGARY)

Dénes König was born in Budapest in 1884, a son of the eminent mathematician Gyula (Julius) König. He published his first mathematics paper at the age of 14, and this was followed soon after by two short books on mathematical recreations, written for a wide readership.[1] After studying for two years at the University of Budapest, he spent five semesters at the University of Göttingen. There he attended Hermann Minkowski's lectures on analysis situs (see Interlude A) and subsequently published a short paper on the four color problem.[2]

After the completion of his doctorate in 1907 with a dissertation on geometry, König joined the staff of the Technical University of Budapest, where he spent the rest of his life. There he lectured on a range of topics, including set theory, real numbers and functions, analysis situs, and graph theory. An excellent teacher with a cheerful and outgoing personality, he inspired a generation of Hungarian mathematicians, some

Dénes König (1884–1944)

of whom (such as Paul Erdős and Paul Turán) would become well-known figures in the world of mathematics. Further information about König's life and mathematical activities can be found in an article by Tibor Gallai, his only doctoral student.[3]

König contributed to several of the topics that we met in Interlude A. In 1911, he wrote two papers on the genus of graphs drawn on orientable surfaces,[4] building on the earlier contributions of Percy Heawood and Lothar Heffter. He was also inspired by Julius Petersen's 1891 paper on the decomposition of regular graphs into 1-factors (or perfect matchings), as we shall see.

In 1914, König attended a Congress of Mathematics and Philosophy in Paris, where he presented some results on bipartite graphs. A graph is *bipartite* if its set of vertices can be divided into two sets so that every edge joins a vertex of one set to a vertex of the other, and König proved that this is the case if and only if every cycle in the graph has an even number of edges. A *complete bipartite graph* is a bipartite graph in which every vertex in each set is joined to every vertex of the other set; if the sets have r and s vertices, the complete bipartite graph is denoted by $K_{r,s}$.

König's results developed some work on matrices by the German algebraist Georg Frobenius and appeared in an important paper, "On graphs and their applications in determinant theory and set theory", that was published in 1916 in Hungarian and German.[5] In its opening section, König investigated factorization in bipartite graphs, by coloring all the edges so that adjacent edges are colored differently and looking at "alternating paths" (paths of edges in two alternating colors). He obtained the following result:

> If each vertex of a bipartite graph meets at most k edges, then we can assign one of k colors to each edge of the graph, so that no two edges that meet have the same color.

In modern terminology, this states that the chromatic index of a bipartite graph is the largest vertex-degree.

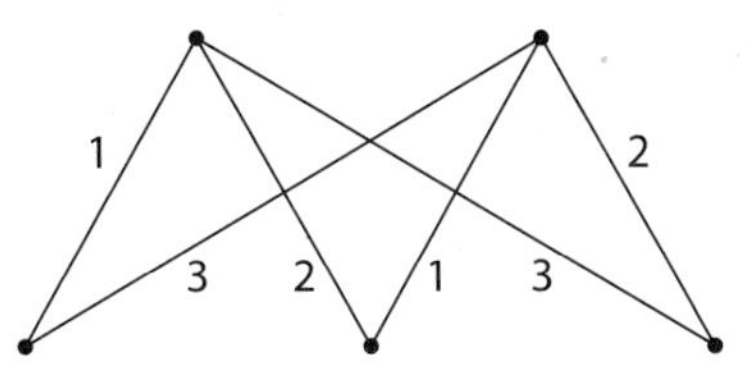

A coloring of the edges of the bipartite graph $K_{2,3}$.

He deduced, because all the edges in each color form a 1-factor in the graph, that

> Every bipartite graph that is regular of degree k splits into k 1-factors—that is, it is 1-factorizable

—in particular,

> Every regular bipartite graph has a 1-factor.

König also presented this remarkable result in the language of matching:

> If, at a dancing party, every man is acquainted with k women, and every woman is acquainted with k men, then we can match them up in such a way that every pair is acquainted.

This was the earliest result in what came to be known as *matching theory*, an area of study that would feature prominently in the ensuing years, as we shall see.

Two years later, following his earlier papers on the drawing of graphs on orientable surfaces, König wrote the monograph *The Elements of Analysis Situs*, in which he carefully presented results on this topic.[6] It was the first book on topological graph theory to be published anywhere, and it provided an early opportunity for many of his readers to learn about the classification of surfaces.

One feature of König's work that distinguishes him from most other writers on graph theory is that, arising from his lifelong interest in set theory, a topic he had shared with his father, he was particularly fascinated by infinite graphs. In important papers from 1926 and 1927, he presented his *infinity lemma*, which provides the means for extending results on finite graphs to infinite ones;[7] for example:

> If, on a given surface, every map with a finite number of regions can be colored with k colors, then so can maps with infinitely many regions.

A *minimax theorem* in graph theory is one in which the maximum value of one graph parameter is equal to the minimum value of another. Konig's most important contribution to graph theory is often considered to be the following minimax theorem, which he proved in 1931, again using alternating paths;[8] here, a *matching* is a set of edges, no two of which meet at a common vertex:

> *König's theorem*: In any bipartite graph, the maximum size of a matching is equal to the minimum number of vertices that collectively meet all the edges.

A generalization of this result to weighted graphs (graphs with numbers associated with their edges) was then developed by his compatriot Jenő (or Eugene) Egerváry.[9] This proved to be central to assignment problems, as we shall see in Chapter 5.

Another extension of König's theorem, in this case to infinite graphs, was discovered in 1932 by the 19-year-old Hungarian mathematician Paul Erdős, while attending König's graph theory class, and was published by König.[10] Later, we shall meet other examples of minimax theorems, such as Menger's theorem and the max-flow min-cut theorem for capacitated networks.

In several of his papers, König applied his results to matrices and determinants, following the work of Frobenius. For example, his minimax theorem can be restated as:

> In any matrix the maximum number of non-zero entries, no two of which appear in the same line, is equal to the minimum number of lines that contain all the non-zero entries.

Here, the lines (rows and columns) of the matrix correspond to the two sets of vertices in the bipartite graph.

In the late 1920s, König wrote what has often been described as "the first textbook on graph theory", his *Theorie der endlichen und unendlichen*

Graphen (Theory of Finite and Infinite Graphs).[11] This thoroughly researched and carefully written work was published in 1936, and became influential throughout Europe, with no other book on the subject appearing until the late 1950s. An English translation of König's book did not become available until 1990.[12]

Written in German, König's book had thirteen chapters and dealt with many of the now standard topics, such as Eulerian and Hamiltonian graphs, trees and forests, directed graphs, factorization, and minimax theorems. As its title indicates, substantial attention was paid to infinite graphs, and there were also sections on other topics such as labyrinths, automorphism groups, and game theory.

THEORIE DER ENDLICHEN
UND UNENDLICHEN GRAPHEN

KOMBINATORISCHE TOPOLOGIE DER STRECKENKOMPLEXE

VON

DÉNES KÖNIG

A. O. PROFESSOR AN DER KGL. UNG. JOSEFS-UNIVERSITÄT FÜR TECHNISCHE UND WIRTSCHAFTSWISSENSCHAFTEN IN BUDAPEST

MIT 107 FIGUREN

LEIPZIG 1936

AKADEMISCHE VERLAGSGESELLSCHAFT M. B. H.

Theory of Finite
and
Infinite Graphs

Dénes König

Birkhäuser
Boston • Basel • Berlin

König's *Theorie der endlichen und unendlichen Graphen* (Theory of Finite and Infinite Graphs).

König's life came to a tragic end, and he never lived to see the success of his book or the subsequent rise in interest in graph theory. During World War II, he had been greatly involved in helping persecuted mathematicians, but on October 15, 1944, Germany invaded Hungary. Coming from a Jewish family, he then became highly vulnerable, and four days later he committed suicide.

ALFRED ERRERA (BELGIUM)

Alfred Errera spent most of his career at the University of Brussels. A lone Belgian voice in graph theory, and map coloring in particular, he published many papers on the subject. In the 1920s, he was working along similar lines as his American counterparts, with whom he was in correspondence, and his writings included many references to them.

In 1921, Errera was awarded a doctorate for his thesis, *Du Coloriage des Cartes et de Quelques Questions d'Analysis Situs*, which was soon published in Brussels and Paris.[13] Although it was mainly on the map color theorem, it included a simpler proof of Petersen's theorem (see Interlude A), which he also published separately.[14] Also featured in his thesis was a graph that he used to explain the error in Kempe's attempted proof of the four color theorem; Percy Heawood, another mathematician with whom Errera corresponded, had presented a different example in his paper of 1890.

Errera was also interested in other problems of a recreational nature. A well-known example is the *utilities problem*, mentioned by the famous

DU COLORIAGE DES CARTES
ET DE QUELQUES QUESTIONS
D'ANALYSIS SITUS

THÈSE

PRÉSENTÉE

A LA FACULTÉ DES SCIENCES
DE L'UNIVERSITÉ LIBRE DE BRUXELLES
POUR OBTENIR LE GRADE DE DOCTEUR SPÉCIAL
EN SCIENCES PHYSIQUES ET MATHÉMATIQUES

PAR

ALFRED ERRERA

DOCTEUR EN SCIENCES PHYSIQUES ET MATHÉMATIQUES

SOUTENUE DEVANT LA FACULTÉ LE 3 DÉCEMBRE 1920
PRÉSENTÉE EN SÉANCE PUBLIQUE LE 20 DÉCEMBRE 1920

IXELLES
IMPRIMERIE G. BOTHY
RUE DE LA CONCORDE 22
1921

Alfred Errera (1886–1960) and his doctoral thesis.

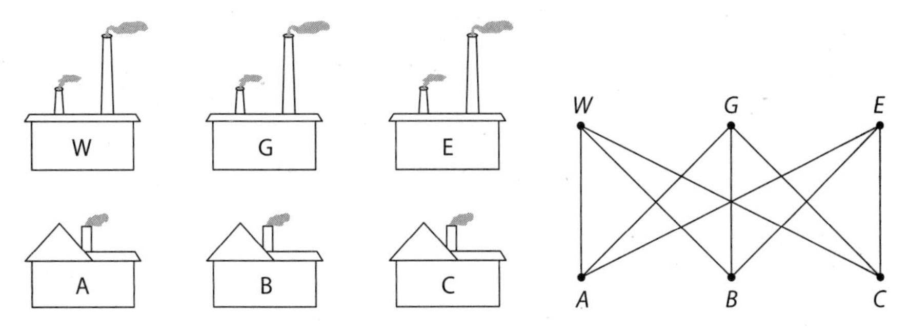

The utilities problem and the complete bipartite graph $K_{3,3}$.

American puzzler, Sam Loyd, around 1900. It was described by his English contemporary, Henry E. Dudeney, in the following terms:[15]

> For some quite unknown reason I have lately received an extraordinary number of letters (four of them from the United States) respecting the ancient puzzle that I have called "Water, Gas, and Electricity." It is much older than electric lighting, or even gas, but the new dress brings it up to date. The puzzle is to lay on water, gas, and electricity, from W, G, and E, to each of three houses, A, B, and C, without any pipe crossing another. Take your pencil and draw lines showing how this should be done. You will soon find yourself landed in difficulties.

In graphical terms, the utilities problem asks us to draw the complete bipartite graph $K_{3,3}$ in the plane without any edges crossing. It is easy to insert eight edges, but it is impossible to draw all nine, as a simple argument involving Euler's formula shows.

In 1923, Errera wrote a short paper in which he generalized the problem to an arbitrary number of utilities and houses.[16] His main result was as follows:

> If the points $A_1, A_2, \ldots, A_u$: $B_1, B_2, \ldots, B_v$ lie in a plane, then exactly $2u + 2v - 4$ of the uv edges A_iB_j, and not more, can be drawn without any two of the edges crossing each other in the plane ($u > 1, v > 1$).

Over the next few years, Errera returned to the coloring of maps, writing five further papers. In 1924, he gave a brief presentation[17] at the International Congress of Mathematicians, held in Toronto, at which he outlined the history of the four color problem from Cayley and Kempe to Heawood and Petersen, and summarized the conclusions of Birkhoff's paper on reducibility and Franklin's paper of 1922. He also listed all the reducible configurations that were known at the time and recalled Franklin's proof that an irreducible map must have at least 26 regions. Three years later, he expanded his presentation into "A historical exposition of the four color problem".[18] This comprehensive article included most of the results that had been proved up to that time, including a discussion of Petersen's theorem and Reynolds's recent result that an irreducible map must have at least 28 regions.

In other papers, Errera contributed to the study of reducibility. In 1925, his main conclusion was that a map whose regions are all pentagons and hexagons is reducible.[19] It follows from the Counting theorem (see Chapter 3) that every irreducible map must have at least one region with seven or more sides, and that it must contain at least thirteen pentagons.

Like his English correspondent, Percy Heawood, Alfred Errera continued to write papers until the 1950s on topics that were mainly related to the four color problem.[20]

ANDRÉ SAINTE-LAGUË (FRANCE)

Jean André Sainte-Laguë (pronounced "sant-lagoo") is mainly remembered for his work in political theory, with a method that he presented at the age of 28 for allocating political seats following an election. A man of wide interests, he was a prolific writer in both the academic and popular spheres.

Born in southwestern France in 1882, he spent part of his childhood in Haiti before returning to his home country. Following military service, he entered the prestigious École Normale Supérieure at the age of 20, after which he spent several years in high school teaching.

Sainte-Laguë joined the infantry during World War I and was awarded military honors, but was later wounded and invalided out to the Department of Invention and the laboratories of the École Normale Supérieure. There he studied the aeronautics of artillery shells and the flight of birds, and began his investigations into the mathematics of graphs and networks.

Among Sainte-Laguë's lifelong interests was recreational mathematics, a fascination that he had probably acquired from Édouard Lucas's four-volume *Récreations Mathématiques* and other books of games and puzzles. The enjoyment of such problems as Euler's Königsberg bridges and Hamilton's Icosian game seems to have led him toward working for a doctoral degree in graph theory, which he was awarded in 1924.

In 1926, Sainte-Laguë published his thesis in book form, with the title of *Les Réseaux (ou graphes)* (On Networks, or graphs).[21] This is

André Sainte-Laguë (1882–1950) and his monograph on graph theory.

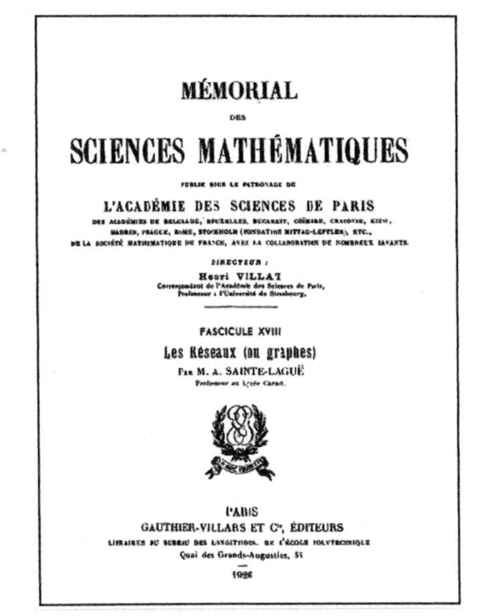

MÉMORIAL
DES
SCIENCES MATHÉMATIQUES
PUBLIÉ SOUS LE PATRONAGE DE
L'ACADÉMIE DES SCIENCES DE PARIS
DES ACADÉMIES DE BELGRADE, BRUXELLES, BUCAREST, COÏMBRE, CRACOVIE, KIEW,
MADRID, PRAGUE, ROME, STOCKHOLM (FONDATION MITTAG-LEFFLER), ETC.,
DE LA SOCIÉTÉ MATHÉMATIQUE DE FRANCE, AVEC LA COLLABORATION DE NOMBREUX SAVANTS.

DIRECTEUR :
Henri VILLAT
Correspondant de l'Académie des Sciences de Paris,
Professeur à l'Université de Strasbourg.

FASCICULE XVIII
Les Réseaux (ou graphes)
Par M. A. SAINTE-LAGUË
Professeur au Lycée Carnot.

PARIS
GAUTHIER-VILLARS ET Cie, ÉDITEURS
LIBRAIRES DU BUREAU DES LONGITUDES, DE L'ÉCOLE POLYTECHNIQUE
Quai des Grands-Augustins, 55

1926

sometimes referred to as "the zeroth book on graph theory"; it precedes by ten years Dénes König's groundbreaking textbook on the subject, which includes several references to his predecessor. Sainte-Laguë's monograph has recently been translated into English as *The Zeroth Book of Graph Theory* and provides us with a useful picture of the topics that were studied at the time.[22]

After some introductory definitions, *Les Réseaux* presents chapters on trees, chains and cycles (including the tracing of mazes), regular graphs, cubic graphs (including Petersen's theorem), tableaux (matrices) and bipartite graphs, and Hamiltonian graphs. His book concludes with two chapters on chessboard problems and an extensive bibliography.

Les Réseaux includes no mention of the drawing of graphs on surfaces or the coloring of maps, but three years later, Sainte-Laguë made up for this with a monograph on the geometry of situation and games, in which these topics were amply covered.[23] This second book mentioned the reducible configurations of Birkhoff, Franklin, Errera, and Reynolds, and concluded with a discussion of many games and recreations, such as the Tower of Hanoi, the game of Fan-Tan (or Nim), the Josephus problem, and Kirkman's 15 schoolgirls problem.

In 1938, André Sainte-Laguë was appointed professor of mathematics for applications at the Conservatoire National des Arts et Métiers. As well as his many publications in abstract and recreational mathematics, he wrote increasingly on the applications of his subject to science and engineering and in the natural world. He also organized the mathematics rooms of the Palace of Discovery for the Paris World's Fair of 1937.

As a highly popular and innovative lecturer on a wide range of topics, Sainte-Laguë employed such aids as the showing of films for his teaching of geometrical ideas. He emphasized an understanding of basic principles, rather than the application of rules, to solve practical problems.

By this time, Sainte-Laguë was an established public figure, having been elected president of the International Confederation of Intellectual Workers; he also became vice president of the National Economic Council. During World War II, he worked for the underground resistance movement and for a short while suffered imprisonment by the German occupying forces. He was made an officer of the Legion of Honour and was decorated with the Croix de Guerre for his patriotic efforts.

KARL MENGER (AUSTRIA)

We have seen that König obtained a "minimax theorem", in which the maximum of one quantity is equal to the minimum of another. His interest in such results arose partly from a theorem of the Austrian mathematician Karl Menger on the connectedness of graphs.

Karl Menger was born in Vienna, the son of the economist Carl Menger. In 1924, he received his doctorate from the University of Vienna for a dissertation on the theory of curves and dimension theory, and became a member of the Vienna Circle, a recently formed discussion group of philosophers and scientists. After two years of teaching in Amsterdam, he returned to Vienna in 1927. He spent the academic year of 1930–31 at Harvard University and the Rice Institute in Houston, Texas. In the mid-1930s he migrated to the United States, where he spent the rest of his life teaching and researching at Notre Dame University in Indiana, and at the Illinois Institute of Technology in Chicago. He worked in several areas of mathematics and, apart from his theorem in graph theory, is remembered mainly for his researches into curves and dimension—and, in particular, for the "Menger sponge", a three-dimensional fractal curve.

One of the most important problems in graph theory is to determine how connected a given graph is. One way of interpreting this problem is to ask how many disjoint paths there are joining two given vertices;

Karl Menger (1902–1985).

another is to ask how many vertices or edges one needs to remove from the graph in order to disconnect it. Such issues arise in a wide range of practical areas, ranging from the connectivity of airline networks to the security of telecommunications networks.

Menger's approach to such problems resulted in a celebrated minimax theorem in a paper that he submitted while investigating the properties of curves; it was published in 1927, during his time in Amsterdam.[24] Here, v and w are two fixed vertices of a finite graph, and the question is to find the maximum number of paths from v to w that have no vertices in common (apart from v and w).

> *Menger's theorem*: Let v and w be non-adjacent vertices in a connected graph. Then the maximum number of vertex-disjoint paths joining v and w is equal to the minimum number of vertices whose removal separates v from w.

It follows that if, for some number k, we can find k vertex-disjoint paths and k vertices that separate v from w, then k must be the *maximum* number of disjoint paths from v to w, and also the *minimum* number of vertices whose removal separates v from w.

There are also many variations of Menger's theorem, such as those for edge-disjoint paths and directed graphs, but these did not appear until later.

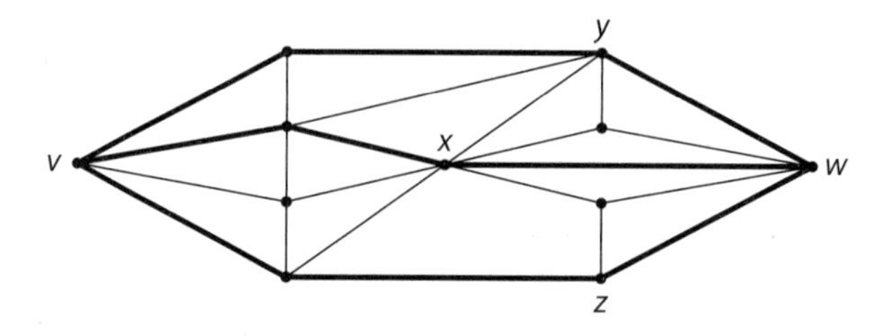

An example illustrating Menger's theorem, with three vertex-disjoint paths from v to w, and three vertices (x, y, and z) that separate v from w.

Unfortunately, Menger's own proof of his theorem contained a serious gap, in that it omitted the special case of bipartite graphs; the first complete proof was given by his German doctoral student, Georg Nöbeling. Dénes König also filled this gap, and Menger then used König's result to present a complete proof in his book on the theory of curves.[25]

Another result that is closely related to Menger's theorem is *Hall's theorem* on selecting representatives of cosets in group theory, which was first proved by Philip Hall of Cambridge (England) in 1935. It is often referred to as the "marriage theorem", after the following reformulation by Paul R. Halmos and Herbert E. Vaughan:[26]

> *Hall's "marriage" theorem*: Suppose that each of a collection of boys is acquainted with a collection of girls. Then each boy can marry one of his acquaintances if and only if, for each number k, every set of k boys is collectively acquainted with at least k girls.

This condition is clearly necessary and also turns out to be sufficient. What is less obvious is that Hall's theorem and Menger's theorem are equivalent—each can be deduced from the other.

KAZIMIERZ KURATOWSKI (POLAND)

Kazimierz (or Casimir) Kuratowski was a Polish mathematician whose main interests lay in point-set topology and related areas. He also made a significant contribution to graph theory in his celebrated theorem on planarity that was published in 1930.

Kuratowski was born in Warsaw in Vistula Land, a part of Poland that was under Russian control. In 1913, following his years at high school,

Kazimierz Kuratowski (1896–1980).

and like other Polish students who did not wish to study in the Russian language, he enrolled in the excellent engineering school at the University of Glasgow, being awarded the class prize in mathematics at the end of his first year.

On arriving home for the summer vacation, he suddenly found himself unable to return to Scotland because of the outbreak of World War I. In 1915, the Russians withdrew from Warsaw and the University of Warsaw was reopened. Kuratowski enrolled as one of its first mathematics students.

By 1921, Kuratowski had completed his studies and was awarded a doctorate for a spectacular dissertation in which he showed how to construct point-set topology from its closure axioms (now sometimes called the "Kuratowski closure axioms"). In 1923, he took up an academic post at the University of Warsaw, but four years later he was appointed to a full professorship at the University of Lwów (now Lviv, in Ukraine), where he stayed until 1934, before returning to Warsaw. It was while he was at Lwów that he carried out his investigations on planar graphs.

A graph G is *planar* if it can be drawn on the plane without edges crossing; any such depiction is a *plane drawing* of G. Such a drawing divides the set of points of the plane not lying on G into *regions*; one of these is of infinite extent and is the *external region*. The number of regions is given by *Euler's formula*:

> For a plane drawing with V vertices, E edges, and R regions, $V-E+R=2$.

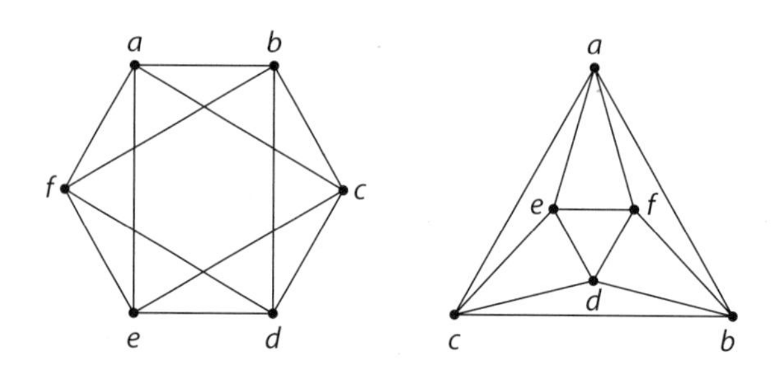

A graph and a plane drawing with eight regions.

Not all graphs are planar—for example, one can show, either directly or by using Euler's formula, that the complete graph K_5 and the complete bipartite graph $K_{3,3}$ (the graph of the utilities problem) are both non-planar.

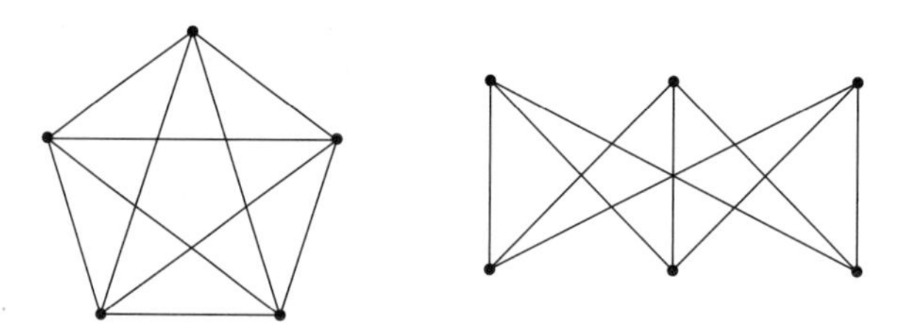

The non-planar graphs K_5 and $K_{3,3}$.

Inserting vertices of degree 2 into the edges of a graph does not affect the planarity of the graph. We say that two graphs are *homeomorphic* if they can be obtained from each other by the insertion or removal of vertices of degree 2.

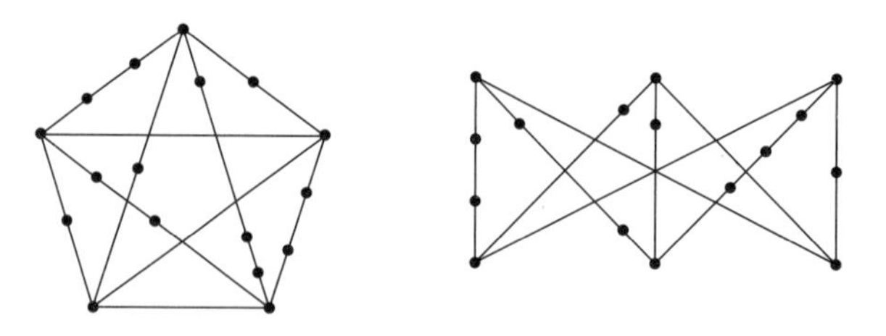

Graphs that are homeomorphic to K_5 and $K_{3,3}$.

In June 1929, Kuratowski announced his results on planar graphs to the Warsaw section of the Polish Mathematical Society. Written from the standpoint of analytic topology, these results were published in 1930.[27] His central theorem was the following:

> *Kuratowski's theorem*: A graph is planar if and only if it has no subgraph that is homeomorphic to K_5 or $K_{3,3}$.

Kuratowski claimed that he had initially believed the only obstruction to planarity to be K_5, and that it was only later when $K_{3,3}$ emerged on the scene:[28]

> I must confess that when I started to think that problem over, I had in mind just one graph. Namely, the graph called now commonly K_5.
>
> But I noticed soon that there is another one which is also irreducibly non-embeddable in the plane. Namely the graph $K_{3,3}$. Now (fortunately for me) [there does] not exist any other irreducible skew graph.
>
> I proved [this] in 1929 in my paper.

Kuratowski was not the only person to be working on this topic. Around the same time, Karl Menger proved the more restricted result that every non-planar *cubic* graph has a subgraph that is homeomorphic to $K_{3,3}$.[29] Two American mathematicians—Orrin Frink of Pennsylvania State University (whom we have already met in connection with Petersen's theorem) and Frink's former colleague Paul A. Smith of Barnard College, New York—independently arrived at the same conclusions as Kuratowski. Frink and Smith prepared a paper for the *Transactions of the American Mathematical Society* and submitted the following abstract to the society's *Bulletin*:[30]

> *Irreducible non-planar graphs*. One of the results of this paper is a simple necessary and sufficient condition that an arbitrary linear graph be mappable on a plane. (Received February 10, 1930.)

As Frink informed one of us in 1974:[31]

> Unfortunately Kuratowski's proof came out in *Fundamenta* [*Mathematicae*] just at that time, and equally unfortunate was the fact that our proof was similar to Kuratowski's. Hence our paper was simply rejected by the *Transactions*.

He later remarked that:[32]

> Kuratowski's proof was actually different from ours, since he did not use the notion of an irreducible non-planar graph, but the two papers were not different enough so that ours could be published.

Claims have also been made that the distinguished Russian topologist Lev Semyonovich Pontryagin obtained a proof of the theorem during the winter of 1927–28, while still a second-year student at Moscow State University; the theorem is now often referred to in Russia as "the Pontryagin–Kuratowski theorem". But because Pontryagin seems never to have communicated or published his proof, it is difficult to judge how valid these claims are; they are analyzed in detail in a historical article by John W. Kennedy, Louis V. Quintas, and Maciej J. Sysło.[33]

In later life, Kuratowski had a most distinguished career. He played a central role in Polish mathematical life, collaborated with such major figures as Stefan Banach, Max Zorn (on "Zorn's lemma"), Stanisłav Ulam, and John von Neumann, and won international awards while acting as a world ambassador for Polish mathematics.

* * * * *

The years around 1930 marked a significant change in graph theory, as the movement from map coloring toward graph theory as a worthy study in its own right gathered momentum. In the next chapter, we shall see how this change in direction was strengthened by the seminal contributions of Hassler Whitney.

Chapter 4
The 1930s

When the distinguished graph-theorist W. T. Tutte provided an introductory commentary for the English edition of Dénes König's *Theory of Finite and Infinite Graphs*, originally published in German in 1936, he observed:[1]

> Low was the prestige of Graph Theory in the Dirty Thirties. It is still remembered, with resentment now shading into amusement, how one mathematician scorned it as "The slums of Topology". It was the so-called science of trivial and amusing problems for children, problems about drawing a geometrical figure in a single sweep of the pencil, problems about threading mazes, and problems about colouring maps and cubes in cute and crazy ways. It was too hastily assumed that the mathematics of amusing problems must be trivial, and that if noticed at all it need not be rigorously established.

Hassler Whitney and Saunders Mac Lane, two of the leading American mathematicians of the 20th century, changed all this with the depth of their discoveries. Early in his career, Whitney made significant contributions to graph theory, with his doctoral thesis in 1932 and several papers on topics ranging from planar graphs and coloring to connectivity and separability. He also invented matroids, a notion that generalizes independence in vector spaces and links the areas of linear algebra and graph theory. Inspired by Whitney's work, Mac Lane wrote three papers on graph theory in the late 1930s, obtaining further conditions for a graph to be planar.

We conclude this chapter by assessing the state of American mathematics in the 1930s, prior to World War II.

HASSLER WHITNEY

Hassler Whitney came from a prominent American family.[2] His namesake was Ferdinand Hassler, the first superintendent of the US Coast Survey, whose granddaughter married Whitney's maternal grandfather,

Hassler Whitney (1907–89).

the astronomer and mathematician Simon Newcomb (whom we met in Chapter 1). Whitney's paternal grandfather was William Dwight Whitney, a professor of ancient languages at Yale, whose eldest brother was Josiah Whitney, the state geologist of California after whom Mount Whitney was named. Whitney's parents were Edward Whitney, an attorney general who was appointed to the New York Supreme Court, and Josepha Newcomb Whitney, an accomplished artist who was active in the women's rights movement and became a member of the Connecticut State Legislature.

Hassler Whitney was born in Cornwall in the Hudson River Valley, about fifty miles north of New York City. He was 3 years old when his father died, and his mother moved her six children to New Haven to live with two sisters-in-law; one of these, Aunt Emily, was an accomplished pianist who greatly influenced his life. Hassler was also close to his eldest sister, Caroline, who encouraged his mathematics and physics projects as a young boy and with whom he later corresponded regularly. When he was 14, his mother took him to Switzerland, where he spent two years at school, learning French in the first year and German in the second. With the Alps so accessible, much of his time was spent in mountain climbing with his elder brother, Roger, and this remained a passion for the rest of his life.

When Hassler was 17, he enrolled at Yale University and majored in physics, taking just one mathematics course (complex variables) in his senior year. Music also became a major part of his life: he played piano, violin, and viola, and his *Fantasie for Orchestra* won the award for the best original composition in his senior year. After completing his bachelor's degree in physics in 1928, he stayed on for a further year to complete a degree in music.

Whitney intended to continue to graduate school in theoretical physics at Harvard, where he had already been accepted. First, however, he spent three weeks in the summer of 1930 visiting Göttingen, whose university was then considered the foremost center for mathematics in the world. While there, he turned to the serious business of preparing for the upcoming semester by reading some extensive notes on the "General Theory of Mathematical Physics" that he had taken during a course at Yale, but things did not go well. As he later recalled:[3]

> I had physics notes to review, which I thought would go quickly; instead I found that I had forgotten most of it, in spite of much recent physics study. Seeing Hilbert–Ackermann, *Grundzuge der Theoretishcen Logik*, in a bookstore, I got it and started working on it, along with George Sauté, a math student from Harvard.

During his short time in Göttingen, he also enjoyed stimulating conversations with Paul Dirac about problems in number theory and with a student from Yale who had enthused about the four color problem:

> So I soon decided that since physics required learning and remembering facts, which I could not do, I would move into mathematics. I have always regretted my quandary, but never regretted my decision.

Back in the United States, he duly arrived at Harvard University, but to study mathematics. Once there, he became obsessed with the four color problem and spent much free time outside class thinking about it. Whitney showed his ideas to George Birkhoff, who had worked on the problem for many years (see Chapter 2). Birkhoff was impressed and became his doctoral advisor.

Whitney's thesis *The Coloring of Graphs* was inspired by Birkhoff's 1912 paper on chromatic polynomials. Birkhoff advised Whitney to submit an immediate summary of his results to the National Academy of Sciences to establish priority, and then to follow this with a full account in the *Annals of Mathematics*.[4]

While Whitney was a graduate student at Harvard, Birkhoff arranged an instructorship for him in mathematics, which he held from 1930 to 1931 and from 1933 to 1935, moving to Princeton University in the intervening years on a fellowship from the National Research Council. It was during his time at Princeton that he wrote most of his papers on graph theory. In 1935, Whitney was appointed to an assistant professorship at Harvard, and promoted to associate professor in 1940 and to full professor in 1946, a position that he held until 1952. He then transferred to the Institute for Advanced Study at Princeton, where he remained for the rest of his life.

Whitney's earliest researches were in graph theory, and in the early 1930s he wrote a dozen papers on the subject. These papers provided a brief, yet extraordinary, contribution to the subject, and ranged over planarity, coloring, combinatorics, and other areas, as we now discover.

Coloring

As we saw in Chapter 2, George Birkhoff had introduced a quantitative approach to the four color problem in 1912, when he defined $P(\lambda)$ to be the number of ways of coloring the regions of a map with at most λ colors, where neighboring regions are assigned different colors. He showed that this function is always a polynomial and derived a general formula for producing its coefficients by means of a difficult proof involving determinants.

By 1930, around the time that Whitney was beginning his graduate studies at Harvard, Birkhoff had once again taken up the four color problem, proving (as we saw in Chapter 2) that

$$P(\lambda) \geq \lambda(\lambda-1)(\lambda-2)(\lambda-3)^{n-3}$$

for any map with $n \geq 3$ regions and for any positive integer λ (except 4). Whitney, who had written up Birkhoff's 1930 paper for him,[5] then applied this same quantitative approach to graphs, rather than maps, using the *dual graph* of the map, as explained by Kempe in his paper of 1879 (see Chapter 1):

> Place a vertex within each region of the map, and join two vertices by an edge whenever the corresponding regions have a common boundary.

In this way, a *coloring* of a graph is an assignment of colors to its vertices, where adjacent vertices are assigned different colors. Coloring the regions of a map is then equivalent to coloring the vertices of its dual graph. Whitney used the notation $M(\lambda)$ for the number of ways of coloring a given graph with at most λ colors; it is now known as the *chromatic polynomial* of the graph.

In 1932, Whitney produced the following paper, which developed the quantitative approach to the four color problem by presenting explicit and efficient procedures for computing the coefficients of a chromatic polynomial, $M(\lambda)$.[6]

Hassler Whitney: *A logical expansion in mathematics* (1932)

Whitney's main achievement in this paper was an expression for the coefficients of the chromatic polynomial, similar to that found by Birkhoff in 1912, although Whitney's approach was far simpler. It was derived from the *principle of inclusion and exclusion* (which he called "a logical expansion"), a counting method that had originated in Abraham De Moivre's *The Doctrine of Chances* of 1718.

In the introduction to his paper, Whitney described the simplest cases of the principle. He took a finite set of objects (his example was the books on a table), where each object either has, or does not have, a given property A (of being red, say). He let n be the total number of objects, $n(A)$ the number with the property A, and $n(\bar{A})$ the number without the property A, so $n(\bar{A}) = n - n(A)$.

He next took a set of n objects with two properties, A_1 and A_2, let $n(A_1 A_2)$ be the number with both properties, and $n(\bar{A}_1 \bar{A}_2)$ be the number with neither property, and observed that

$$n(\bar{A}_1 \bar{A}_2) = n - n(A_1) - n(A_2) + n(A_1 A_2).$$

For three properties the corresponding formula is

$$\begin{aligned} n(\bar{A}_1 \bar{A}_2 \bar{A}_3) = n &- (n(A_1) + n(A_2) + n(A_3)) \\ &+ (n(A_1 A_2) + n(A_1 A_3) + n(A_2 A_3)) - n(A_1 A_2 A_3). \end{aligned}$$

In the second section of the paper (which he called *The Logical Expansion*), Whitney used mathematical induction to prove the corresponding formula for m properties—that is, $n(\bar{A}_1 \bar{A}_2 \cdots \bar{A}_m)$. Commenting that this extension to the general case is quite simple, and that it was well known to logicians of the time, he added that it should also be better known to mathematicians.

To illustrate its usefulness, Whitney applied the principle to three areas of mathematics. He first derived an expression for the number of integers up to a given number that are not divisible by any of a given set of primes. He next calculated the probability that, if a pack of m cards is laid out on a table, and if another pack of m cards is laid out on top of it, then no card of the second pack lies over the same card of the first. This is the *derangement problem* of Pierre Raymond de Montmort and Nicholas Bernoulli from around 1710. The probability is the sum of the first $m+1$ terms in the series expansion of $1/e$.

Finally, Whitney turned to his most significant application. He considered a simple graph G with v vertices. If λ colors are available, then there are λ^v possible colorings of the vertices in total, and he wished to find $M(\lambda)$, the number of these colorings in which no two adjacent vertices have the same color. To do so, for each edge ab of G, he let A_{ab} be the set of colorings in which the vertices a and b are colored alike. The set of admissible colorings is then

$$\bar{A}_{ab}\,\bar{A}_{bd}\cdots\bar{A}_{cf},$$

where the subscripts range over all the edges of G, and so

$$M(\lambda)=n(\bar{A}_{ab}\,\bar{A}_{bd}\cdots\bar{A}_{cf}).$$

Whitney then applied his logical expansion formula. To evaluate a typical term, $n(A_{ab}\,A_{ad}\cdots A_{ce})$, of the logical expansion, he considered the corresponding subgraph H in which all the edges ab, $ad,\ldots,$ ce have ends of the same color; in particular, all the vertices in a single connected component of H are colored the same. So if H has p components, then the value of this term in $M(\lambda)$ is λ^p. Also, if H has s edges, then the sign of the term is $(-1)^s$, and so $(-1)^s\, n(A_{ab}\,A_{ad}\cdots A_{ce})=\lambda^p$. Using Birkhoff's symbol (p, s) for the number of subgraphs with p components and s edges, Whitney noted that the corresponding terms contribute $(-1)^s\,(p, s)\,\lambda^p$ to the chromatic polynomial, and so

$$M(\lambda)=\textstyle\sum_{p,s}(-1)^s\,(p, s)\,\lambda^p.$$

Next, following the ideas of Oswald Veblen (see Chapter 2), Whitney defined the *rank* i and the *nullity* j of a subgraph H with v vertices, s edges, and p components by

$$i=v-p \quad\text{and}\quad j=s-i=s-v+p, \quad\text{so that}\quad p=v-i \quad\text{and}\quad s=i+j.$$

Finally, on putting $(p, s) = m_{v-p, s-v+p} = m_{ij}$, where m_{ij} is the number of subgraphs of G with rank i and nullity j, he obtained his final formula for the chromatic polynomial:

$$M(\lambda) = \sum_{i,j} (-1)^{i+j} m_{ij} \lambda^{v-i} = \sum_i m_i \lambda^{v-i}, \quad \text{where} \quad m_i = \sum_j (-1)^{i+j} m_{ij}.$$

Whitney then proceeded to develop a simple method for determining the coefficients m_i of the chromatic polynomial, in terms of what he called the *broken circuits* of G—that is, cycles with one edge removed: specifically, the number $(-1)^i m_i$ turns out to be the number of subgraphs with i edges which do not contain all the edges of any broken circuit. We illustrate his method with the two examples from his paper.

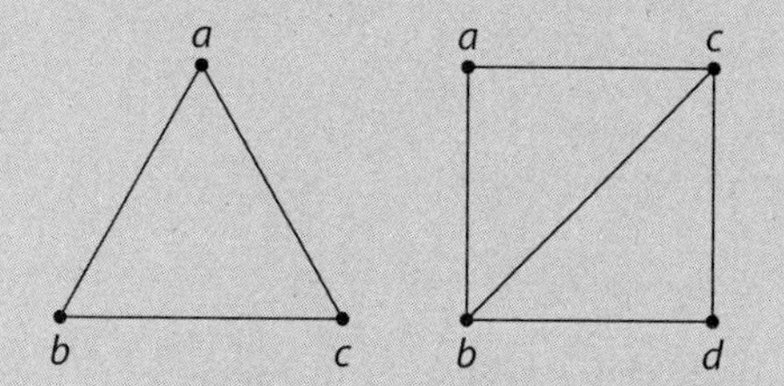

Example 1. Let G be the complete graph K_3, with vertices a, b, c, and with the edges ab, ac, and bc, listed in this definite order. Then there is only one cycle—ab, ac, bc, with the edges listed in the same order—and from this, we form the broken circuit ab, ac by dropping *the last edge*. Then:

only one subgraph has 0 edges, so $m_0 = 1$;
three subgraphs have 1 edge, so $-m_1 = 3$;
three subgraphs have 2 edges, one of which contains the broken circuit, so $m_2 = 2$;
one subgraph has 3 edges which contain the broken circuit, so $-m_3 = 0$.

So, because $v = 3$,

$$M(\lambda) = \lambda^3 - 3\lambda^2 + 2\lambda.$$

This is easily verified, because there are λ ways to color vertex a, $\lambda - 1$ colors left for vertex b, and then $\lambda - 2$ colors left for vertex c, and so

$$M(\lambda) = \lambda(\lambda - 1)(\lambda - 2) = \lambda^3 - 3\lambda^2 + 2\lambda.$$

Example 2. Let G be the complete graph K_4, with one edge removed, with vertices a, b, c, d, and with the edges ab, ac, bc, bd, and cd, listed in this definite order. Then there are three cycles,

$$ab, ac, bc; \quad bc, bd, cd; \quad \text{and} \quad ab, ac, bd, cd$$

(with their edges listed in the same order), and from these, we form the broken circuits

$$ab, ac; \quad bc, bd; \quad \text{and} \quad ab, ac, bd$$

by dropping the last edge from each cycle. Then:

only one subgraph has 0 edges, so $m_0 = 1$;
five subgraphs have 1 edge, so $-m_1 = 5$;
ten subgraphs have 2 edges, two of which are broken circuits, so $m_2 = 8$;
ten subgraphs have 3 edges—but here we can ignore the last broken circuit because it contains the first one, and only four subgraphs contain neither of the first two, so $-m_3 = 4$;
five subgraphs have 4 edges, and each contains a broken circuit, so $m_4 = 0$.

So, because $v = 4$,

$$M(\lambda) = \lambda^4 - 5\lambda^3 + 8\lambda^2 - 4\lambda.$$

Again, this is easily verified, because there are λ ways to color vertex a, $\lambda - 1$ colors left for vertex b, and then $\lambda - 2$ colors left for vertices c and d, and so

$$M(\lambda) = \lambda(\lambda - 1)(\lambda - 2)^2 = \lambda^4 - 5\lambda^3 + 8\lambda^2 - 4\lambda.$$

Whitney made several other significant contributions to the coloring of graphs. In his *Annals of Mathematics* paper “The coloring of graphs”, a revised form of his doctoral thesis, he had obtained further results on the coefficients m_i of the chromatic polynomial $M(\lambda)$—in particular, that they alternate in sign, as in the preceding examples. He also proved that we can obtain these coefficients more efficiently by considering only “non-separable” subgraphs H, instead of the much larger collection of all subgraphs. Here, a connected graph is *non-separable* if it cannot be disconnected by the removal of a single vertex (a *cut-vertex*) and is *separable* otherwise. A maximal non-separable subgraph of a graph is called a *block*.

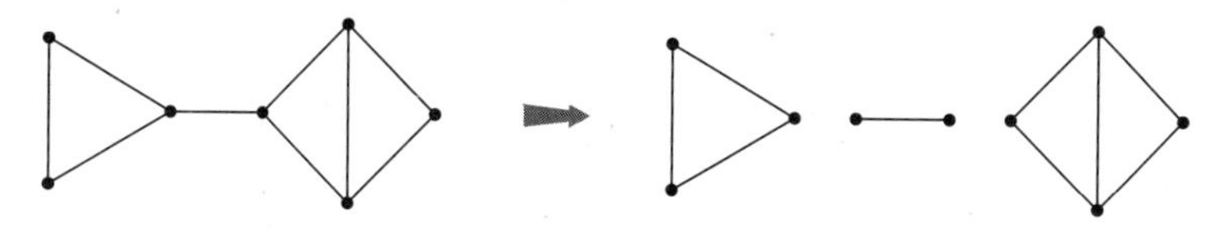

The blocks of a separable graph.

As a curiosity, we note that Whitney's *Annals* paper concluded with a calculation of the chromatic polynomial of the dodecahedron graph:

$$M(\lambda)=\lambda(\lambda-1)(\lambda-2)(\lambda-3)\times(\lambda^8-24\lambda^7+260\lambda^6-1670\lambda^5 +6999\lambda^4-19698\lambda^3+36408\lambda^2-40240\lambda+20170).$$

We have seen that Whitney was fascinated by the four color problem, and we conclude this section with three further results that he obtained on this topic.

The first of these also comes from his *Annals* paper. If G is a planar graph with rank r and nullity n, if G^* is a dual of G, and if m_{ij} and m_{ij}^* are the corresponding coefficients for G and G^*, then $m_{ij}^*=m_{r-j,n-i}$. Now, if $\mathcal{C}$ is the class of graphs for which all the coefficients m_{ij} and m_{ij}^* arise from graphs, then $\mathcal{C}$ includes all the planar graphs. Whitney showed that if

$$\textstyle\sum_{i,j}(-1)^{i+j}m_{ij}\,4^{v-i}>0$$

can be proved for all graphs in $\mathcal{C}$, then the four color theorem follows. As he observed, this condition is stronger than the four color theorem, because there are graphs in $\mathcal{C}$ that are not planar.

In another paper on graph coloring, "A theorem on graphs",[7] communicated to the American Mathematical Society on February 22, 1930, Whitney presented the following remarkable result.

> Given any simple planar graph in which every region has exactly three boundaries (a "triangulation"), but where no other cycle has three edges, there is always a cycle that passes through every vertex—that is, a Hamiltonian cycle.

After deriving the dual form for cubic planar graphs, he deduced that, when trying to solve the four color problem, we can restrict our attention to planar graphs that are Hamiltonian.

According to Whitney, Alfred Errera was intrigued by this paper and corresponded with him about it. Whitney visited him in Belgium in 1931 and 1933, where they "talked a lot about graph theory".[8]

The four color conjecture is usually thought of in geometric terms, but in his last paper on coloring, Whitney presented alternative forms that are more combinatorial in nature.[9] One of these involves sums of numbers.

Consider the sum $a+b+c+d$. We can evaluate it by first inserting parentheses in various ways, such as in the "arranged sums"

$$((a+b)+c)+d, \quad (a+(b+c))+d, \quad (a+b)+(c+d), \quad \text{or} \quad a+(b+(c+d)).$$

We then carry out the additions in order, as directed by the parentheses, listing the "partial sums" that we encounter along the way; for example, the partial sums for $(3+7)+(1+4)$ are the single numbers 3, 7, 1, 4, the pairs $3+7=10$ and $1+4=5$, and the final addition $10+5=15$. Whitney then proved that the four color problem can be restated, as follows:

> If an n-fold sum (any n) is expressed in any two ways as an arranged sum, is it always possible to choose the terms of the sum as integers so that no partial sum of either arranged sum is divisible by 4?

This paper concludes with an estimate by Whitney, along the lines of the earlier ones of Heawood and Franklin, of the tiny "probability" that there is a map with more than N regions that cannot be colored with four colors, when N is sufficiently large.

Planarity and Duality

Several of Hassler Whitney's most important contributions to graph theory involved his investigations into planarity and duality, based on the concepts of rank and nullity that we encountered in his work on coloring. In particular, he sought to replace the geometric idea of duality, as introduced by Kempe (see Chapter 1), with equivalent combinatorial definitions in which geometry does not feature. For ease of reading, we shall write G^* instead of Whitney's G'.

Whitney began by recalling that, if a connected graph G can be mapped on a sphere, then it can be mapped on a plane by stereographic projection, and conversely. Moreover, by rotating the sphere in an appropriate manner, we can arrange for any specified region of the map to be the external region of G. A *geometric dual* G^* is then formed from a plane drawing of G, as follows:

> inside each region of G select a point v^*—these points are the vertices of G^*;
>
> for each edge e of G, draw a line or curve e^* that crosses e (but no other edge of G) and joins the vertices v^* in the regions adjoining e—these lines are the edges of G^*.

It follows that, if a plane drawing of G has V vertices, E edges, and R regions, then G^* has R vertices, E edges, and V regions, and that the dual G^{**} of G^* is simply G. Note that different plane drawings of G can lead to different geometric duals G^*.

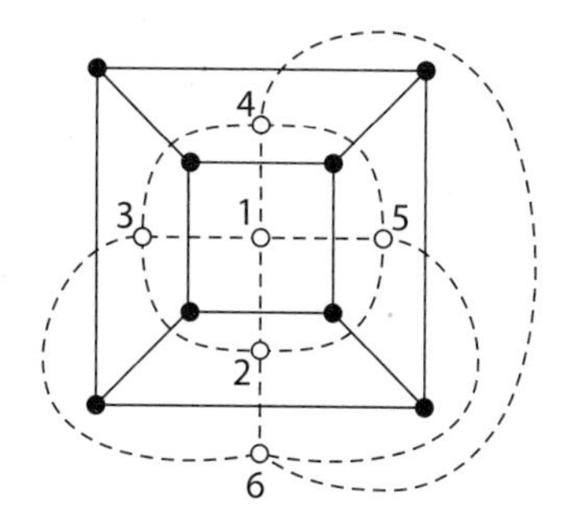

A plane drawing of a planar graph G, and a geometric dual G^*.

On January 14, 1931, Whitney communicated to the American Mathematical Society his results on planarity and duality. Like his paper on the coloring of graphs, these were first published in abstract form by the National Academy of Sciences to establish priority, before appearing in full detail.[10]

Hassler Whitney: *Non-separable and planar graphs* (1932)

In the first part of this paper, Whitney began by defining the *rank* $r(G)$ and *nullity* $n(G)$ of a graph G with V vertices, E edges, and P connected components:

$$r(G) = V - P \quad \text{and} \quad n(G) = E - r(G) = E - V + P.$$

He then derived some simple consequences of these definitions, such as:

- for any graph, the rank and the nullity are non-negative;
- if isolated vertices are added or removed, the rank and the nullity are unchanged;
- if an edge ab is added, where a and b are in the same component of G, the rank remains unchanged and the nullity increases by 1;
- if an edge ab is added, where a and b are in different components of G, the nullity remains unchanged and the rank increases by 1.

This was followed by several substantial results on the ranks and nullities of separable and non-separable graphs.

In the second part of this paper, Whitney turned his attention to duality. Given a graph G, he defined the graph G^* to be a *dual* of G if there is a one–one correspondence between the set of edges of G and the set of edges of G^* with the property that, if H is any subgraph of G with the same vertex-set as G, then the corresponding subgraph H^* of G^* satisfies

$$r(G^*) = r(\hat{H}^*) + n(H),$$

where $\hat{H}^*$ is the complement of H^* in G^* (the graph obtained from G^* by removing the edges of H^*).

This abstract form of dual graph is combinatorial in nature, rather than geometric, because there is no mention of the graphs being drawn on the plane, the sphere, or any other surface. We shall refer to it as a *combinatorial dual* (sometimes called a *Whitney dual*) to distinguish it from the geometric dual described earlier.

After presenting his definition, Whitney proved that:

If G^* is a dual of G, then $r(G^*) = n(G)$ and $n(G^*) = r(G)$.

The first result follows on putting $H = G$, so that $\hat{H}^*$ is an empty graph and $r(\hat{H}^*) = 0$.

The second result follows from the first, and from the equalities

$$r(G^*) + n(G^*) = |E(G^*)| = |E(G)| = r(G) + n(G).$$

He also proved that:

If G^* is a dual of G, then G is a dual of G^*.

This is because

$$\begin{aligned} r(H) + n(\hat{H}^*) &= (|E(H)| - n(H)) + (|E(\hat{H}^*)| - r(\hat{H}^*)) \\ &= |E(H)| + |E(\hat{H})| - r(G^*) \text{ (because } |E(\hat{H}^*)| = |E(\hat{H})|) \\ &= |E(G)| - n(G) = r(G). \end{aligned}$$

Note that this is a general result which makes no reference to the geometric situation for connected planar graphs that we described earlier.

Whitney's main achievement in this paper was to prove that

A graph is planar if and only if it has a combinatorial dual.

This result provides a purely combinatorial (non-geometric) characterization of planarity, and was a major contribution to the study of planar graphs and to graph theory in general.

To prove it, Whitney began by showing the necessity of the condition, by mapping a planar graph G of nullity n onto the surface of a sphere, and proving that G divides the surface of the sphere into $n+1$ regions.

He next constructed the graph G^*, by placing a new vertex in each region of G, and crossing each edge of G with an edge of G^* that joins those vertices of G^* lying on either side of it. This produces a one–one correspondence between the edges of G and those of G^*.

Whitney then had to prove that the graphs G and G^* satisfied his definition of duality. To do so, he first built up G, one edge at a time, removing the corresponding edge from G^* whenever he added an edge to G. He showed that:

each time the nullity of G increases by 1 when an edge is added, removing the corresponding edge of G^* decreases the number of components of G^* by 1;
each time the nullity of G remains the same when an edge is added, the number of components of G^* remains the same.

Now let H be a subgraph of G, and let $\hat{H}^*$ be the complement of the corresponding subgraph of G^*. Once again, Whitney built up H one edge at a time, removing the corresponding edges of G^* at the same time as he formed $\hat{H}^*$. By the preceding results, the increase in the number of connected components in forming $\hat{H}^*$ is equal to the nullity of H—that is,

$$P(\hat{H}^*) - P(G^*) = n(H).$$

But, because G^* and $\hat{H}^*$ contain the same vertices,

$$r(\hat{H}^*) = V(\hat{H}^*) - P(\hat{H}^*) = V(G^*) - P(\hat{H}^*) \text{ and } r(G^*) = V(G^*) - P(G^*).$$

It follows that

$$r(G^*) - r(\hat{H}^*) = n(H)$$

—that is, G^* is a combinatorial dual of G, as desired.

To prove the sufficiency of the condition—that is, if a graph has a combinatorial dual, then it is planar—Whitney claimed that it was enough to show this for non-separable graphs. To do so, he first proved that if a separable graph G has a dual, then its blocks also have duals—thus, its blocks are planar, and so G is planar. Sufficiency was therefore a consequence of the following theorem, which he then proceeded to prove:

> Let the non-separable graph G have a dual G^*. Then we can map G and G^* together onto the surface of a sphere, so that corresponding edges in G and G^* cross each other, no other two edges cross each other, and inside each region of one graph there is just one vertex of the other graph.

This completed the proof of Whitney's main theorem.

This paper of Whitney's concluded with a proof that neither K_5 nor $K_{3,3}$ has a combinatorial dual. It follows that Kuratowski's characterization in Interlude B—that a graph is planar if and only if it has no subgraph homeomorphic to K_5 or $K_{3,3}$—can be used to give an alternative proof that a graph that has a combinatorial dual is planar. In a subsequent paper, entitled simply "Planar graphs",[11] Whitney reversed the argument, by proving that a graph that contains no subgraph homeomorphic to K_5 or $K_{3,3}$ has a combinatorial dual, thereby obtaining a new (and less geometric) proof of Kuratowski's theorem.

There is an addition to this story, in that the paper just cited included a further non-geometric characterization of planar graphs. It involves the cycles and cutsets of a connected graph, where a *cutset* is a set of edges whose removal splits the graph into two pieces, and is minimal with respect to this property. Whitney's alternative characterization was as follows:

> Two connected graphs G and G^* are duals if there is a one-one correspondence between their edge-sets with the property that a set of edges in G forms a cycle of G if and only if the corresponding set of edges of G^* forms a cutset of G^*.

We shall call G and G^* *abstract duals* of each other.

To indicate why this result is true, we take a plane drawing of G and let C be a cycle. Then C encloses one or more finite faces of G, and so surrounds a non-empty set S of vertices of its geometric dual G^*. Those edges of G^* that cross the edges of C then form a cutset of G^*, whose removal disconnects G^* into two subgraphs, one with vertex-set S and the other containing the remaining vertices. The converse argument is similar.

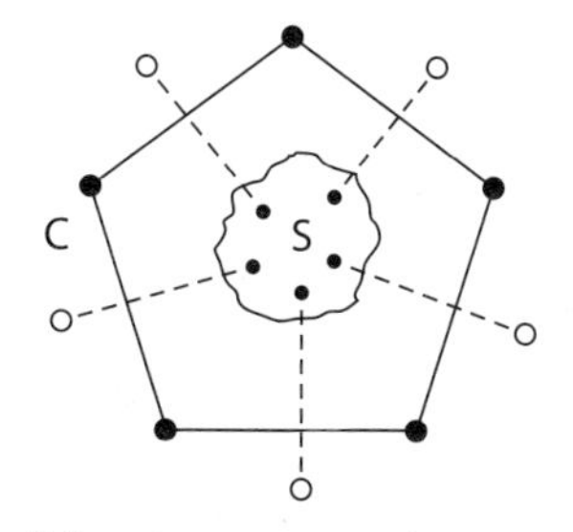

A cycle (solid) and a corresponding cutset (dotted).

To prove that abstract duals are the same as combinatorial duals, Whitney argued as follows. Suppose that cutsets in G^* correspond to cycles in G. Let H be a subgraph of G with nullity $n(H)$, and let $\hat{H}^*$ be the complement of the corresponding subgraph of G^*. We can then form H and $\hat{H}^*$ simultaneously, by starting with no edges in G and every edge in G^*, and adding edges one at a time to H, while removing the corresponding edges from G^*. At each stage, the nullity of the resulting subgraph of G increases if and only if the last edge added forms a cycle with the edges already present, and the rank of the subgraph of G^* decreases if and only if the last edge removed forms a cutset with the edges already removed. So, because cycles in G correspond to cutsets in G^*, the nullity of the first subgraph increases if and only if the rank of the second one decreases. It follows that

$$r(G^*) = r(\hat{H}^*) + n(H),$$

and so G^* is a combinatorial dual of G. Whitney proved the converse result in a similar way.

Matroids

In September 1934, Hassler Whitney presented to the American Mathematical Society his investigations into matroids, which were published in the following year.[12] Having spent several years working on graphs, he had noticed similarities between the ideas of rank and independence in graph theory and those of dimension and linear independence in vector spaces. We have seen how Whitney used the complementary concepts of rank and nullity as a foundation for his studies, and in his groundbreaking paper on matroids he went further, motivated by the following correspondences between algebraic vectors and the edges of a graph:

subsets of a vector space	$\leftrightarrow$	subsets of edges in a graph
dimension of a span of vectors	$\leftrightarrow$	rank of a subgraph
linearly independent vectors	$\leftrightarrow$	cycle-free sets of edges
basis of a vector space	$\leftrightarrow$	spanning forest in a graph
minimal dependent set	$\leftrightarrow$	cycle of a graph

Whitney first axiomatized the properties of a rank function, mirroring results on dimension in a vector space and the rank function in graph theory, and then presented three equivalent axiom systems for independent sets, bases, and cycles. The resulting structure, called a *matroid*, has proved to be of fundamental importance, not only in graph theory but also in other branches of mathematics.

Hassler Whitney: *On the abstract properties of linear dependence* (1935)

A *matroid* M consists of a non-empty finite set E of elements, where for each subset S of E, there is an integer $r(S)$, called the *rank* of S, satisfying the following conditions:

(R1) $r(\varnothing)=0$;
(R2) for any subset S, and any element e not in S,

$$r(S \cup (e)) = r(S) + k, \text{ where } k=0 \text{ or } 1;$$

(R3) for any subset S, and any elements e and f not in S,

$$\text{if } r(S \cup (e)) = r(S \cup (f)) = r(S), \text{ then } r(S \cup (e,f)) = r(S).$$

It follows that:

for each subset S of E, $0 \le r(S) \le |S|$;
if S is a subset of T, then $r(S) \le r(T)$;
for all subsets S and T of E, $r(S \cup T) + r(S \cap T) \le r(S) + r(T)$.

These last three results can be taken as alternative axioms.

Whitney called a subset S *independent* if $r(S) = |S|$, and *dependent* otherwise. He then defined a matroid in terms of its independent sets, by requiring that:

(I1) any subset of an independent set is independent;
(I2) if S and T are independent sets with $|S| = p$ and $|T| = p+1$, then there is an element e that is in T but not in S for which $S \cup (e)$ is independent.

For a graph, the independent sets are the sets of edges that contain no cycles.

By repeating property (I2) as often as necessary, we eventually obtain a maximal independent set, which he called a *base* of the matroid. Whitney defined a matroid in terms of its bases, by requiring that:

(B1) no proper subset of a base is a base;
(B2) (*exchange axiom*) if B_1 and B_2 are bases, and if e is an element of B_1, then there is an element f in B_2 for which $(B_1 - (e)) \cup (f)$ is a base.

For a connected graph, the bases are its spanning trees.

By successively replacing elements of B_1 by those in B_2, as guaranteed by the exchange axiom, Whitney proved that

any two bases have the same number of elements.

Note that a subset is independent if it is contained in a base, and that a base is a set with rank $r(B) = |B| = r(E)$.

Whitney's last definition of a matroid was in terms of its minimal dependent sets, called *circuits*. We can define a matroid in terms of these by requiring that:

(C1) no proper subset of a circuit is a circuit;
(C2) if C_1 and C_2 are circuits, each containing an element e, then there is a circuit in $C_1 \cup C_2$ that does not contain e.

Now, if G is a graph, we can define a matroid by taking as its circuits the cycles of G, and we call this the *cycle matroid* of G, denoted by $M(G)$. But we could alternatively have taken as circuits the (minimal) *cutsets* of G, because these also satisfy the circuit axioms, and we call this matroid the *cutset matroid* of G, denoted by $M^*(G)$. To see the connection between these two matroids, we show how the idea of duality can be extended to matroids.

Given any matroid M on a set E, we can construct another matroid on E, called its *dual matroid* M^*, by taking as its bases the complements of the bases of M—so if B is a base of M, then $E-B$ is a base of M^*. It follows that every matroid has a unique dual (in contrast to the duality of planar graphs), and that if M^* is a dual of M, then M is a dual of M^*.

But we can say more. We recall that two connected graphs G and G^* are abstract duals of each other if there is a one–one correspondence between their edge-sets so that a set of edges in G forms a cycle of G if and only if the corresponding set of edges in G^* forms a cutset of G^*. It can be proved that, for any graph G, the circuits of the dual of the cycle matroid $M(G)$ are precisely the cutsets of G. It follows from this that

> The cycle and cutset matroids $M(G)$ and $M^*(G)$ are duals of each other.

Moreover, if we now calculate the rank function r^* of M^*, it turns out to be a restatement of the formula $r(G^*)=r(\hat{H}^*)+n(H)$ for the combinatorial dual of G. It follows that matroid duality is the natural setting for both the abstract and the combinatorial duals of a connected graph.

Whitney's seminal work on matroids was largely ignored for over twenty years, even though Saunders Mac Lane wrote an article in 1936 on connections between matroids and projective geometry,[13] B. L. van der Waerden formalized the ideas of algebraic and linear dependence in 1937 in his classic textbook, *Moderne Algebra*,[14] and in 1942 Richard Rado developed the connections between matroids (under the name of "independence systems") and transversal theory.[15] The situation changed dramatically in 1958, when W. T. Tutte presented a characterization of those matroids that arise from graphs (see Chapter 5).

Other Topics

Whitney's papers on graph theory represented an important advance in the subject. Indeed, as Tutte later pronounced:[16]

> The graph-theoretical papers of Hassler Whitney, published in 1931–1933, would have made an excellent textbook in English had they been collected and published as such.

We have already chronicled Whitney's work on graph coloring, planarity, and matroids, and we now turn to a few other topics that he investigated in the early 1930s.

Isomorphism and line graphs

One of Whitney's earliest papers, communicated to the American Mathematical Society on February 28, 1931, was "Congruent graphs and the connectivity of graphs".[17] It was a remarkable achievement, in that it contributed to two very different areas of graph theory.

Two graphs, G and G', are said to be *isomorphic* if there is a one–one correspondence between their vertex-sets, so that two vertices are adjacent in G if and only if the corresponding vertices are adjacent in G'. (Whitney referred to such graphs as *congruent*.) In graph theory, isomorphic graphs are usually regarded as indistinguishable, even though drawings of them may look dissimilar; for example, the three drawings of the Petersen graph in Interlude A look markedly different.

Whitney also investigated other forms of isomorphism. For example, G and G' are *edge-isomorphic* if there is a one–one correspondence between their edge-sets, so that if two edges of G meet at a common vertex, then so do the corresponding edges of G'. It is easy to check that if the graphs G and G' are isomorphic, then they are also edge-isomorphic, but is the converse result necessarily true? Whitney proved that this is always the case, except when one graph is the complete graph K_3 and the other is the complete bipartite graph $K_{1,3}$.

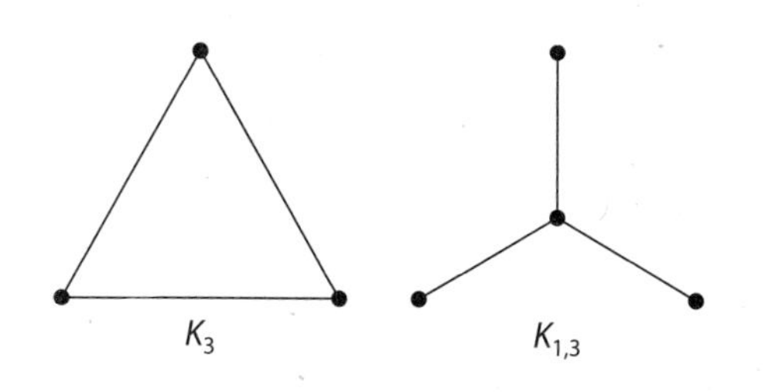

The graphs K_3 and $K_{1,3}$.

Another way of expressing this result (not explicitly mentioned by Whitney) is to define the *line graph* $L(G)$ of a connected graph G to be the graph whose vertices correspond to the edges of G, with two vertices of $L(G)$ adjacent if and only if the corresponding edges of G meet. It follows from Whitney's result that, if $L(G)$ is isomorphic to $L(G')$, then G is isomorphic to G', except when G and G' are K_3 and $K_{1,3}$.

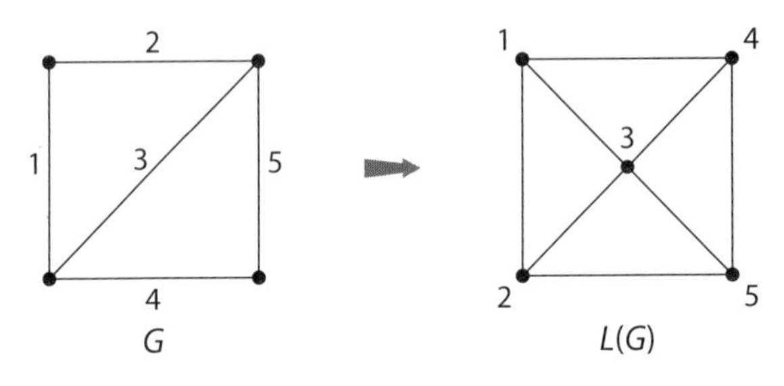

A graph G and its line graph $L(G)$.

Further types of isomorphism were developed in two later papers, both published in 1933. In the first of these, "2-isomorphic graphs",[18] Whitney investigated the relationships between two graphs in which the cycles of one graph are in a one–one correspondence with the cycles of the other. In the second paper, "On the classification of graphs",[19] he built on an investigation by Ronald Foster of the Bell Telephone Company, "Geometrical circuits of electrical networks".[20] Whitney had corresponded with Foster on the subject, and his paper began:

> R. M. Foster has given an enumeration of graphs, for use in electrical theory. He used two distinct methods, classifying the graphs according to their nullity, and according to their rank. In either case, only a certain class of graphs is listed; the remaining graphs are easily constructed from these. In the present paper we give theorems sufficient to put the first method of classification on a firm foundation.

To do so, Whitney considered certain operations on graphs, and he explored the various types of homeomorphism and isomorphism which can exist between two graphs that are related by combinations of these operations. He concluded his paper by outlining a method for constructing graphs with given nullities.

The connectivity of graphs

In the second half of his paper "Congruent graphs and the connectivity of graphs", Whitney devised a measure of how "joined up" a given connected graph is. A connected graph is "1-connected". It is "2-connected" if it is a block—that is, it remains connected if a single vertex is deleted (with its incident edges). Whitney extended this idea by defining a connected graph to be *n-connected* if it remains connected whenever up to $n-1$ vertices are deleted. If a graph is n-connected, then it is also k-connected for all values of k that are less than n.

In Interlude B we met Menger's theorem, which states that if v and w are non-adjacent vertices in a finite connected graph, then the maximum number of vertex-disjoint paths joining v and w is equal to the minimum number of vertices whose removal separates v from w. Whitney expressed this minimax result as follows:

> A graph with at least $n+1$ vertices is n-connected if and only if any two vertices are joined by at least n paths, no two of which have any other vertices in common.

He further defined the *connectivity* of a graph to be n if it is n-connected but not $(n-1)$-connected. It is the minimum number of vertices whose deletion leaves a disconnected (or trivial) graph.

The final section of Whitney's paper concerned the duals of planar graphs. We have mentioned that different plane drawings of the same planar graph G can lead to different geometric duals. Whitney proved, among other results, that if G is further assumed to be 3-connected, then it has a unique dual.

TOPOLOGICAL INVARIANTS OF GRAPHS

In Interlude B, we defined two graphs to be *homeomorphic* if they can be obtained from each other by the insertion or removal of vertices of degree 2. In his paper, "A set of topological invariants for graphs",[21] Whitney investigated those properties of a graph that are the same for homeomorphic graphs. This paper was presented to the American Mathematical Society on December 28, 1931, but was not published until 1933.

A simple example of a topological invariant for a graph is the *number of components*, which remains unchanged when vertices of degree 2 are inserted or removed. Another example is the *nullity*, which is $E-V+P$ for a graph with E edges, V vertices, and P components, because if a vertex of degree 2 is inserted or removed, then E and V both increase or decrease by 1.

A more substantial example arises from Whitney's work on chromatic polynomials. In his 1932 paper, "A logical expansion in mathematics", which we described earlier, he defined m_{ij} to be the number of subgraphs with rank i and nullity j. Whitney now asserted:

> Given the table of the numbers m_{ij} for a graph G, if we sum over the elements in each row with alternating signs, we get the coefficients m_i of the polynomial $M(\lambda)$ for the number of ways of coloring G

in λ colors. But if we sum over the columns, we get a set of numbers p_i. If G is of rank r and nullity n, then these numbers are

$$p_i = \sum_j (-1)^{i+j} m_{r-j,n-i}.$$

Whitney proved that if G is a planar graph, then the numbers p_i are topological invariants for G. For, if G has a dual G^*, and if the corresponding numbers are m^*_{ij}, then $m^*_{ij} = m_{r-j,n-i}$, and so

$$m^*_i = \sum_j (-1)^{i+j} m^*_{i,j} = \sum_j (-1)^{i+j} m_{r-j,n-i} = p_i.$$

It follows that if G has a dual G^*, then the numbers p_i are the coefficients of the chromatic polynomial of G^*. Moreover, if G is planar, then these coefficients p_i remain unchanged when vertices of degree 2 are inserted or removed, and so they are indeed topological invariants for G.

Whitney proved further that, just as $(-1)^i m_i$ is the number of subgraphs with i edges that do not contain all the edges of any broken circuit, so $(-1)^i p_i$ is the number of subgraphs with i edges that do not contain all the edges of any broken cutset.

Whitney's Later Life

After Whitney's early work in graph theory, he turned his attention to other topics, which we outline briefly.[22]

- It was well known that any continuous function on a closed subset of $\mathbb{R}^n$ can be extended to a continuous function on all of $\mathbb{R}^n$. In an important paper, Whitney extended this result from continuous functions to k-differentiable (or even infinitely differentiable) ones.
- Based on this result, Whitney proved his *strong embedding theorem*, that any n-dimensional differentiable manifold can be embedded in $\mathbb{R}^{2n}$. This result is optimal, because the Klein bottle is a 2-dimensional manifold that can be embedded in $\mathbb{R}^4$, but not in $\mathbb{R}^3$.
- In the 1930s, homology groups were found to be insufficiently structured to distinguish between topologically different manifolds. The solution was to replace them by "cohomology rings", but a suitable ring product proved difficult to find. Working with Eduard Čech, Whitney (then aged just 28) proposed the *cup product*, and showed that different manifolds

always have different cohomology rings, even when their homology groups are the same.
- In the 1950s, by now at the Institute for Advanced Study, Whitney was engrossed in the topology of singular spaces and the singularities of smooth maps. This work led to catastrophe theory (founded by René Thom), and eventually to non-linear dynamics and chaos theory, fields that are still very active today.
- In 1965, Whitney published a major paper on *stratifications*, in which he described how to decompose a non-smooth manifold into smooth manifolds. He proposed that a "good stratification" needed to satisfy two conditions, and it was later proved that these *Whitney conditions* provide just the right definition for a stratification.

During his long and successful career, Whitney's achievements were recognized by honorary doctorates and memberships in learned societies. He presented the American Mathematical Society's Colloquium Lectures in 1946 on the *Topology of Smooth Manifolds*, and became vice president of the society in 1948–49. He also received a number of awards, including the National Medal of Science, conferred on him in 1976 by President Jimmy Carter.

By the 1970s, much to the puzzlement and consternation of some colleagues, Whitney had abandoned his career as a research mathematician and embarked upon a second one. Motivated by his discovery of the appalling way in which his own children were being taught mathematics, he devoted the last two decades of his life to improving mathematics education for schoolchildren.

SAUNDERS MAC LANE

Saunders Mac Lane is best known for co-founding category theory with Samuel Eilenberg. Born in Norwich, Connecticut, he was christened Leslie Saunders MacLane, but his parents came to dislike "Leslie" and it fell into disuse. Later, he began to insert a space in his surname because his wife found it difficult to type it without one. His grandfather was a church minister, but was forced out of the church because he believed in evolution. His father also was a minister, and his mother taught English, Latin, and mathematics. He was the oldest of three brothers (a sister died as a baby), and his brother Gerald became a mathematics professor at Rice University and Purdue University.

Saunders Mac Lane (1909–2005).

Mac Lane enrolled at Yale University in 1926, where he studied mathematics and physics as a double major, and graduated with a bachelor's degree and the best grade point average yet recorded at Yale. He began his graduate work at the University of Chicago, where he studied with E. H. Moore, Leonard Dickson, G. A. Bliss, and the philosopher Mortimer Adler, gaining his master's degree in 1931. With Moore's encouragement, he was awarded a fellowship at the University of Göttingen. While there, he studied with Emmy Noether before receiving his doctorate in 1934 for the thesis *Abbreviated Proofs in the Logical Calculus*, under the supervision of Hermann Weyl and Paul Bernays.

From 1934 to 1936, Saunders Mac Lane was the Peirce Instructor of Mathematics at Harvard University. Here, he met the new assistant professor, Hassler Whitney, and published three papers that extended Whitney's work. The first of these, "Some unique separation theorems for graphs",[23] was published in 1935 and explored some methods of Whitney and R. M. Foster on separating graphs by chains. In the concluding paragraph, Mac Lane observed that his techniques could also be used to study the separation of graphs by cycles.

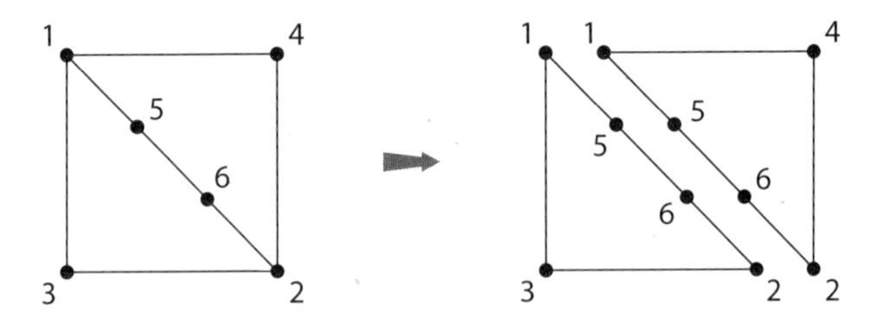

Separating a graph by the chain 1-5-6-2.

He did this in his next paper, "A combinatorial condition for planar graphs",[24] where he provided a new condition for planarity. After recalling Kuratowski's theorem and Whitney's combinatorial dual, Mac Lane proved that

> A combinatorial graph is planar if and only if it contains a complete set of cycles with the property that no edge appears in more than two of these cycles.

Here, a set of cycles is *complete* if every cycle in the graph can be written as a sum (modulo 2) of cycles in the set—that is, each edge of the cycle appears in an odd number of these cycles.

Mac Lane's third paper, "A structural characterization of planar combinatorial graphs",[25] also dealt with planar graphs and cycles. The paper began by recalling the planarity results of Kuratowski, Whitney, and himself, and the condition that he developed was the following:

> A set of cycles $C_1, \ldots, C_m$ in a non-separable graph G is the set of complementary domain boundaries of a planar map of G if and only if each edge of G is contained in exactly two of the cycles C, while the cycles $C_1, \ldots, C_{m-1}$ form a complete independent set of cycles in G, mod 2.

This implies that a non-separable graph is planar if and only if it contains a set of cycles with the property that each edge of the graph lies in exactly two of these cycles.

For two years, Saunders Mac Lane taught at Cornell University and the University of Chicago, before joining the mathematics faculty at Harvard from 1938 to 1947. In 1944–45 he was involved in war work, directing the Applied Mathematics Group at Columbia University. He returned to the University of Chicago as professor of mathematics in 1947 and became the mathematics department's chair in 1952, serving in this capacity for six years. During his teaching career, Mac Lane supervised forty-two doctoral students, including David Eisenbud, Irving Kaplansky, and John Thompson.

Mac Lane was the recipient of many honors. Elected to the National Academy of Sciences in 1949, he later served as its vice president, also becoming president of both the Mathematical Association of America (MAA) and the American Mathematical Society (AMS). He was awarded the MAA's Chauvenet Prize and Distinguished Service Award, and the AMS's Steele Prize, and in 1989 he received the National Medal of Science, America's highest award for scientific achievement.

Saunders Mac Lane authored highly regarded books on a wide variety of mathematical subjects. In 1941, he collaborated with Garrett Birkhoff on *A Survey of Modern Algebra*, an influential text that introduced modern abstract algebra to generations of undergraduates, while his 1972 book, *Categories for the Working Mathematician*, remains the definitive introduction to category theory.

ACADEMIC LIFE IN THE 1930s

The stock market collapse in 1929 led to a worldwide economic downturn, known as the Great Depression, and hard times for all. In particular, the United States and Canada witnessed many business failures that led to high levels of unemployment and privation. The jobless figures rose to 25 percent at a time of no unemployment benefits and few pension plans.

As in other countries, the situation affected academics in varying degrees.[26] In the early 1930s, the salaries of full professors became subject to reductions of 10–15 percent, although this coincided with a general lowering of the cost of living, so those at the higher end of the academic ladder remained relatively well off. But this was not the case for those on the lower rungs, or for those with recently acquired degrees, as too many people with mathematics doctorates found themselves chasing too few jobs. State and central governments drastically reduced their funding for establishments of higher learning, while many colleges and universities were forced to terminate the contracts of junior staff, and there were instances of increased teaching workloads to cover for released staff.

In the 1930s, some mathematicians with new doctoral degrees were unable to gain academic positions related to their own research interests. Others took positions as high school teachers or in industry that was unrelated to their research, or indeed to mathematics in general, and a few even became unemployed. Even forsaking colleges and universities in order to obtain high school positions was not without its problems. Many heads of mathematics departments in high schools were wary of bringing in someone with greater mathematical knowledge and understanding than themselves, but who might not teach well. Even when mathematicians were fortunate enough to obtain teaching positions at a college, university, or high school, it was unlikely that any increases in salary would be awarded, unless they were attached to promotions that could take years to achieve.

Although World War I and the Great Depression undoubtedly had many negative effects on academic life in the 1930s, there were positive developments as well. It was precisely during the Depression years that major academic developments took place at a number of institutions, such as Princeton, Berkeley, Duke University, the University of Virginia, and many of the land-grant universities and elsewhere.

Immigration

In the 1930s, the events in Europe, and particularly those that followed Hitler becoming chancellor of Germany in 1933, had a considerable impact on the world at large. For those who were employed in American higher education, these happenings triggered an exodus of people from German-speaking countries because of persecution. Many Jewish mathematicians migrated to America, where some were subjected to prejudice and anti-Semitism, but all were soon to be caught up in a new global conflict. Most of them actively and proudly assisted their new country and its allies in defeating German and Japanese aggression.

Throughout its history, the United States has been a haven for the oppressed of other countries. It was therefore unsurprising that large numbers of Americans were moved to help those who were subject to persecution in German-speaking countries. These included many academics who became active in the efforts to help individual scholars who were losing their livelihoods for being non-Aryans or for activities that were politically unacceptable to the Nazis.

An Academic Assistance Council had already been formed in Britain, and this influenced efforts that were being made in the United States. An American organization, the Emergency Committee for Displaced German Scholars (later, Displaced Foreign Scholars), was formed to plan the immigration and absorption of these displaced scholars. Indeed, by the conclusion of World War II, up to 120 mathematicians dismissed from their posts by the Nazis had entered the United States, and many remained permanently.

It was no easy task that these agencies had set for themselves, as America was still recovering from the stock market collapse and the subsequent economic depression. Some in the academic world believed that an influx of foreign scholars would deprive homegrown talented young people from obtaining suitable positions. Additionally, there were feelings of nationalism and anti-Semitism to contend with.

But foreign mathematicians benefited, not least through mathematical practitioners in the various organizations—for example, both the president and the head of the natural science program of the Rockefeller Foundation were mathematicians.[27] In addition, many older foreign mathematicians came from European families that had migrated to America only one or two generations earlier, and were of the age to remember the days when American scholars had traveled to Europe for postgraduate study. By so doing, these Americans had established close academic and personal relationships with European universities, many of which had been in Germany, the leading center for mathematics.

As an example of the effects of the Great Depression, it was reported by Edward R. Murrow, the second-in-command of the Emergency Committee, that by October of 1933, 2000 teachers (out of 27,000) had been dismissed from the faculties of 240 American institutions of higher learning.[28] The Emergency Committee and the Rockefeller Foundation decided to use their funds to aid scholarships, while endeavoring not to displace existing faculty members, or to act in such a way as to encourage anti-Semitism or resentment of incoming foreigners. Even with these intentions, their actions were not without critics within the American education system, with some native scholars resentful that the influx of foreigners would prevent young homegrown scholars from progressing up the education ladder.

One of the foremost mathematicians to be actively involved in procuring American entry for displaced scholars from Nazi persecution was Oswald Veblen. Among these immigrants were three notable academics from Göttingen:

> Richard Courant, who was placed on leave of absence by the Nazis in 1933, traveled to the United States via England in 1934, and later became director of the Institute of Mathematical Sciences at New York University;
>
> Otto Neugebauer, who arrived via Denmark, and became professor of the history of mathematics at Brown University in Providence;
>
> Hermann Weyl, who moved to the Institute of Advanced Study in 1933, where he remained until his retirement in 1952.

Insisting on fair play and opportunity for everyone, Veblen had declared:[29]

> One of the greatest dangers . . . is the timid attitude which is taken by most of the scientific people who deal with these questions.

But Veblen himself was anything but timid. In 1943, he declined to fill in the entry for "race" on a form associated with his war work at the army's ordnance Aberdeen Proving Ground; he had, after all, been a major in ordnance, working on ballistics research during World War I. He wrote to the secretary of war, protesting that the question was invidious and of the type that would have been the norm in Germany. In 1946, he again refused to sign a form that waived the right to strike at Aberdeen. A few years later, during Joseph McCarthy's infamous witch-hunt period, it was suggested that Veblen was a Communist and should be denied a passport. But he was not a Communist—rather, he claimed to be an old-fashioned liberal.

Veblen became a member of the Emergency Committee at its foundation and provided detailed information on each possible immigrant. Along with Hermann Weyl, he ran a placement bureau for displaced mathematicians until the end of the war. As described by Nathan Reingold:[30]

> In Veblen's papers in the Library of Congress are lists of names with headings such as scholarship, personality, adaptability and teaching ability. When information about a person was incomplete in the United States, Veblen wrote to European colleagues.

In 1933, the American Mathematical Society formed a committee to cooperate with the Emergency Committee. Veblen was one of the three people appointed to this new committee.

The agreed-upon policy of the two organizations that helped the dispossessed and politically unacceptable foreign academics to enter America was that the most eminent should be placed in institutions with research capabilities. Veblen, Weyl, and others were soon placing the less renowned refugees in those universities, colleges, and junior colleges that would take them.

But this unofficial policy was controversial and did not find support in some quarters—and particularly at Harvard University, which endeavored to raise alternative funds for aiding refugees in an attempt to ease the Emergency Committee aside and regain control of its own faculty appointments. George Birkhoff, in particular, was deeply concerned about the employment opportunities for homegrown mathematicians at a time when immigrant scholars seemed to be gaining an unfair advantage. To avert a major crisis within the mathematics community, Veblen sought an agreement with Birkhoff, and on May 24, 1939, Harlow Shapley, head of the Harvard College Observatory, wrote:[31]

> When Veblen and Birkhoff were in my office the other day, it was agreed that the distribution of these first-rate and second-rate men among smaller American institutions would in the long run be very advantageous, providing

> at the same time we defended not too feebly the inherent right of our own graduate students.

This agreement was probably reached because both of them, and others in the American Mathematical Society, believed that the United States would soon be at war again. Once the war was over, America could feel reasonably proud that, despite all the differing opinions and the many arguments and deals, the nation had again become a haven for the oppressed when it was required.

A New Publication

Prior to 1933 and the rise of the Nazi Party, Germany had been the leading center for mathematics and had published the most respected of reviewing and abstracting vehicles, the *Zentralblatt für Mathematik*. The journal's editor was Otto Neugebauer, who was not Jewish, but who held views that were politically unacceptable to the new German regime; he was forced to flee to Denmark in 1934. Four years later, Tullio Levi-Civita, an Italian mathematician, was dismissed from his professorship and removed from the board of *Zentralbatt* for political reasons. A number of resignations swiftly followed, including those of G. H. Hardy, Harald Bohr, and Oswald Veblen. The journal's management also decided that refugee mathematicians and Russians should be barred from being collaborators in, and referees for, its published articles. The resulting reaction in America and elsewhere was one of great indignation that the independence of scientific internationalism had been violated, and that the worldwide body of mathematicians had been insulted and its integrity impugned.

As a result of the actions of *Zentralblatt*'s governing board and the resignations of advisory members, Veblen urged that American mathematicians should found a new review journal—a suggestion that he had made some fifteen years earlier, even though he recognized that the mathematics community was then unprepared for taking on such a task. But by 1938, Veblen believed that such a move was now possible in America. The number of research mathematicians had considerably increased over the previous two decades, including the recent immigrants, and America was rapidly becoming the world center for mathematical activity.

In December 1938, the American Mathematical Society formed a committee to discuss the idea, and all the different factions of the American mathematics community joined in the deliberations. These covered

Announcing

Mathematical Reviews

an international journal to abstract and review current mathematical literature

This new journal will report upon the mathematical literature of the world as fast as it appears, and will be noteworthy for two important innovations:

The subscription price has been set at a figure which makes it possible for each individual mathematician to have MATHEMATICAL REVIEWS on his desk. Although the price—$6.50 to members of supporting organizations—is admittedly below the cost of publication, it has been kept low so that MATHEMATICAL REVIEWS can function as an instrument of widespread mathematical culture and not merely as a tool of research in the large centers.

A microfilm service operating in conjunction with MATHEMATICAL REVIEWS will place at the disposal of every subscriber the current mathematical literature of the world, rather than mere abstracts of it. Microfilm copies of any article reviewed in MATHEMATICAL REVIEWS will be available at very nominal cost. For those without equipment for reading microfilm, photoprints will be available at cost. No matter where one lives, he can now scan the literature as it appears—through the pages of MATHEMATICAL REVIEWS—and can also have copies of those papers which are of special interest to him.

$6.50 per year to members of sponsoring organizations

$13.00 per year to others

Mathematical Reviews

AMERICAN MATHEMATICAL SOCIETY, *Brown University*, Providence, Rhode Island, U. S. A.

Mathematical Reviews

SPONSORED BY

The American Mathematical Society &
The Mathematical Association of America

January, 1940 TABLE OF CONTENTS **Vol. 1, No. 1**

Announcing *Mathematical Reviews*, 1940.

political, religious, and racial questions, and considered financial security and international cooperation. Veblen used his considerable powers of persuasion to obtain a $65,000 grant from the Carnegie Corporation. After much debate, with many conflicting opinions—not least a pro-German stance by Harvard's mathematics department (including Birkhoff)—the society's council decisively voted on May 25, 1939, in favor of the journal. Veblen was appointed chairman of the committee entrusted to organize and launch the new publication, which was named *Mathematical Reviews*. The first issue appeared in January 1940.

Anti-Semitism

Like other forms of prejudice, anti-Semitism has existed for millennia, as it has waxed and waned over time and moved geographically around the globe. Between the two world wars, it was to be found within American society and, in particular, by some mathematicians at senior institutions of higher education. Prior to a general understanding of Hitler's "final solution" and the evidence of the Holocaust, anti-Semitism had often been seen, not as a mortal sin, but simply as ugly and petty minded.

It is sometimes difficult to distinguish between nationalistic feelings and anti-Semitism; some expressions of national sentiment have no hidden anti-Semitic intent, whereas other nationalistic comments against foreigners mask prejudice. Although America's history had been built on immigration, some newcomers expressed an antipathy toward other foreigners, especially those also wishing to settle in the United States. Because many of the immigrants fleeing Nazi Germany were of Jewish extraction, it was possible to hide anti-Semitic prejudice behind concerns that the incoming academics would prevent homegrown postgraduates from securing appropriate positions.

The 1920s and 1930s witnessed many examples of university faculties operating an anti-Jewish policy, and those which allowed employment for a Jewish scholar rarely hired a second one. However, most members of mathematical departments showed no anti-Semitism and positively worked to undermine the actions of their colleagues who did.

One prominent figure who has been accused of anti-Semitic tendencies was George Birkhoff, and as one of America's two leading mathematicians, his words and actions have been remembered, especially as the other, Oswald Veblen, held opposing views. In 1934, for example, Birkhoff was initially against Solomon Lefschetz's becoming the first Jewish president of the American Mathematical Society, as a letter from him to R.G.D. Richardson, the society's secretary, declared:[32]

> I have a feeling that Lefschetz will be likely to be less pleasant even than he had been, in that from now on he will try to work strongly and positively for his own race. They are exceedingly confident of their own power and influence in the good old USA. The real hope in our mathematical situation is that we will be able to be fair to our own kind . . . He will get very cocky, very racial and use the Annals [of Mathematics] as a good deal of racial perquisite. The racial interests will get deeper as Einstein's and all of them do.

At the AMS's semicentennial meeting in 1938, Birkhoff presented a historical survey of mathematics in America. In his lecture, he discussed foreign-born mathematicians, particularly the recent immigrant scholars, and although his list of names included some who were neither German nor Jewish, many in his audience considered his views anti-Semitic. Long after the occasion, this lecture still caused much heated debate, and even Einstein apparently described him as "one of the world's greatest academic anti-Semites".[33] However, in his article "The migration of European mathematicians to America",[34] Professor Lipman Bers opined that, "even the great mathematician G. D. Birkhoff was not free from anti-Jewish prejudices", but then added:

> Besides, people are complicated. The same Birkhoff who could toss off an anti-Semitic remark in a private letter, did not let his racial prejudices interfere with his evaluations of other peoples' scientific work. The late complex analyst Wladimir Seidel, who graduated from Harvard and later taught there as a Benjamin Peirce Instructor, told me about a phone call made by Birkhoff to a departmental chairman. "I know you hesitate to appoint the man I recommended because he is a Jew. Who do you think you are, Harvard? Appoint Seidel, or you will never get a Harvard Ph.D. on your faculty." Seidel was duly appointed.

Others who have written on this subject in support of Birkhoff have included Saunders Mac Lane, to whom we give the last word:[35]

> It is my view that the 1930s tension between placing refugees and helping young Americans came to a reasonable balance of these interests—and that the differing views of Veblen and Birkhoff served to help the balance.

* * * * *

Like their predecessors, Veblen, Birkhoff, and Franklin, Hassler Whitney and Saunders Mac Lane both had long and successful careers as they contributed to the development of graph theory in America and to mathematics worldwide. On many occasions, the careers and academic interests of these five mathematicians overlapped: Veblen and Birkhoff coincided as graduates of E. H. Moore at Chicago, and then as colleagues at Princeton, as they were the doctoral supervisors of Franklin and Whitney. Moore provided academic encouragement to Mac Lane; and Birkhoff, Whitney, and Mac Lane converged at Harvard in the late 1930s.

These mathematicians all provided academic support to their upcoming colleagues, and the following chapters include notable collaborations between Birkhoff and D. C. Lewis on chromatic polynomials, and between Whitney and W. T. Tutte on the four color problem, as graph theory increasingly established itself on the mathematical scene.

Chapter 5
The 1940s and 1950s

Although many mathematicians in North America and Europe served their countries by contributing to the war effort, research activity continued, and among the American scholars who contributed to graph theory in the 1940s were George Birkhoff, Daniel C. Lewis, Arthur Bernhart, Isidore Kagno, Richard Otter, and Claude Shannon. In this chapter, we outline their achievements, and also describe the notable contributions of Bill Tutte, an Englishman who settled in Canada in 1948, and of Frank Harary, a major figure in the development of graph theory from the 1950s onward. We also survey the graph algorithms that were being developed during and after the war as the computer age began.

WORLD WAR II

The Second World War began with Nazi Germany's attack on Poland in September 1939 and the fall of France to Germany in June 1940. With great public support, President Roosevelt placed the substantial resources of the United States behind the British, but isolationism was a major political force in America at the time. The entry of the United States into World War II seemed far from inevitable, but on December 7, 1941, Japan launched a devastating surprise attack against the US naval and air installations at Pearl Harbor in Hawaii. The United States declared war on Japan and, within days, on Germany and Italy also. A once politically divided nation had finally unified and entered this global conflict. Meanwhile, Canada had entered the conflict in stages, declaring war on Germany in 1939, then on Italy in 1940, and finally on Japan in 1941.

As in World War I, most patriotic citizens willingly contributed their skills to winning this war, with several hundred American mathematicians among them. Some became uniformed combatants as enlisted and civilian staff in such units as the army, the navy, and the navy's Bureau of Aeronautics and Bureau of Ordnance. Others went into industry to work on war-related projects or remained in their academic posts while providing much needed training programs. A number were recruited into cryptanalysis, and also into the Manhattan Project, which had been set up to produce nuclear weapons under the direction of the physicist Robert Oppenheimer at New Mexico's Los Alamos National Laboratory.

Prior to World War II, applied mathematics had not featured highly in the mathematics departments of North American universities and colleges, its development being left mainly to engineers and physicists; this state of affairs had arisen back in 1917 and continued. But the time had come for all those involved in the war effort to tackle whatever duties were assigned to them. Most of these tasks lay within the realm of applied mathematics, with a number of graph theorists rising to the challenge.

Foremost among these was W. T. Tutte. In 1941, he joined the now famous British Government Code and Cypher School at Bletchley Park, and played a major role in decrypting communications enciphered by the Lorenz cipher, as we shall see. For this achievement, he was inducted as an Officer of the Order of Canada in 2001, for "one of the greatest intellectual feats of World War II".

W. T. (Bill) Tutte (1917–2002).

As we saw in Chapter 4, the American mathematician who contributed most to the war effort was Oswald Veblen, who returned to the Ballistics Research Laboratory at Maryland's Aberdeen Proving Ground where he had been assigned during World War I. Veblen took responsibility for recruiting mathematicians, but was sometimes confronted by universities or colleges that were reluctant to release faculty members for the war effort. In such cases, he regularly arranged for foreign-born substitutes who could not work for the American military; one such substitute was Gerhard Karl Kalisch, Veblen's assistant at Princeton's Institute for Advanced Study. In this way, the influx of talented mathematicians from abroad helped to maintain the nation's continuing need for education in mathematics, while also helping to provide the military with mathematicians whose specialized abilities were needed for war work.

In 1942, an Applied Mathematics Panel was formed as a division of the National Defense Research Committee. Headquartered in Manhattan, it was responsible for a variety of research projects, most notably at Princeton and Columbia universities. At the latter institution, the Applied Mathematics Group was directed first by E. J. Moulton and then by Saunders Mac Lane, whose work involved applying differential equations to a particular problem known as the "fire control problem" for air-to-air rockets.

The relentless bombing of London (known as the "Blitzkrieg") began in 1940 and had spread to other cities in Britain. Waves of German planes had met with almost no resistance, because of the British gunners' inability to aim their anti-aircraft weapons at rapidly moving targets. In 1942, Mac Lane recruited Hassler Whitney to join his Applied Mathematics Group to work on this fire control problem. Specifically, Whitney's task was to design a gunsight that could accurately aim a gun at another plane, by making calculations that were based on such data as the range to the target, its angular velocity, and the flight time of the projectile, while the gunner kept the image of the enemy plane within a target circle on a screen. Whitney quickly worked out the relevant differential equations and suggested improvements. In a test of his ideas on two particular flight simulators, one simulator employed the usual aiming method and shot down only two enemy planes, while Whitney's simulator successfully hit them all.

Like all wars, World War II was a period of enormous innovation. Many of the inventions that were developed during the war, such as the atomic bomb, jet engines, and radar, have dramatically changed our lives forever. Others were far less momentous, but still remain with us today. For example, when the United States joined the war, a plant in Newark, New Jersey, began the production of M&M's candies, to be sold exclusively to the military as a convenient way to provide troops with chocolate.

There were also tremendous innovations in mathematics as a result of the war. The field of cryptography underwent enormous development through the advent of electronic computing and the decryption work at Bletchley Park. George Birkhoff had recognized the potential of calculating devices as aids to mathematics and—with funds from a bequest to Harvard University and considerable assistance from IBM—he financed the installation in 1944 of a computer, the Harvard Mark I, that had been designed by the physicist and computer pioneer Howard Aiken. Its success led the US Navy to order three more machines, with improved specification for use in naval laboratories.

GRAPH THEORISTS OF THE 1940s

In this section, we meet five American mathematicians who contributed to the development of graph theory during the 1940s. It was in this decade that the areas of interest began to diverge more widely, from the chromatic polynomials and map coloring of earlier years to automorphism groups, trees, and the coloring of wires in an electrical network.

Daniel C. Lewis

In the early 1940s, George Birkhoff collaborated with D. C. Lewis in an attempt to systematize the previous work on map coloring and to offer further results. Their alliance resulted in a lengthy paper, "Chromatic polynomials",[1] which Lewis presented to the American Mathematical Society on August 23, 1946, two years after Birkhoff's death.

Daniel Clark Lewis was born in New Jersey in 1904 and graduated from Haverford College near Philadelphia. He then transferred to Harvard University, where he obtained his doctoral degree in 1932 for a thesis on differential equations under Birkhoff's supervision. In 1934–35 he attended Princeton's recently founded Institute for Advanced Study. Following the award of a National Research Council Fellowship, he held teaching posts at Cornell University and the University of New Hampshire. From 1943 to 1945, he worked at Columbia University in its war research establishment, after which he was appointed professor of applied mathematics at Johns Hopkins University, retiring in 1971. His main research interests continued to be in differential equations, where he discovered and developed the theory of "autosynartetic" solutions, and generalized the results of Henri Poincaré on periodic solutions of ordinary differential equations.

While working together in 1942, Birkhoff and Lewis received a letter from Clarence Reynolds, claiming a solution of the four color problem that used some of their joint work.[2] Lewis, concerned, wrote to Birkhoff:

> If Reynolds really has solved the problem (I remain sceptical until I have chance to see what he has actually done), don't you think it would be a good thing for us to publish immediately the part of our work on which he based his solutions?

But Reynolds's "proof" was unconvincing, and Lewis told him that if his work were to be published, then he and Birkhoff wished to take no credit for results on which it was based. Reynolds apologetically replied to Birkhoff:

> Referring to my two communications concerning the four color problem which have been sent to you this summer, you will please carefully place the second one in your waste basket. Yes, I have burned my fingers! . . .
> I do sincerely apologise for sending you my last letter.
>
> With a very red face, Clarence N. Reynolds.

One can empathize with him for having to swallow his pride before such a world-renowned mathematician.

In 1942, George Birkhoff and his wife visited Mexico and South America as goodwill ambassadors, cooperating in the efforts of Nelson Rockefeller, coordinator of Inter-American Affairs, to promote solidarity against Hitler. In August, in a letter to Birkhoff about their joint paper, Lewis (who was responsible for its writing and production) commented that he had "so far written some 90 pages", predicting that the final paper would be "about 75 printed pages"; it eventually came to 97 pages. Relying on Birkhoff only for minor suggestions and comments, Lewis eagerly awaited Birkhoff's return to the United States so that the draft paper could be reviewed. But although the manuscript was essentially complete by mid-1943, it was not until November of 1945 that the American Mathematical Society finally received it.

G. D. Birkhoff and D. C. Lewis: *Chromatic polynomials* (1946)

Birkhoff and Lewis's paper opened with a lengthy introduction, relating it to earlier writings on the subject, and succinctly set the scene by separating the two "quite different types of investigation", the quantitative and the qualitative.

The paper had six chapters. Chapter I, on "First principles", was essentially quantitative. It reviewed the basic properties of chromatic polynomials, focusing in particular on reduction formulas for the chromatic polynomial of a given map in terms of those of simpler maps. The main result of this chapter was:

> Let T be an m-gon in a map P_n of n regions. Let $\Pi_{n-k}(\lambda)$ denote the sum of the chromatic polynomials associated with the submaps obtained by erasing just k boundaries of T. Then
>
> $$P_n(\lambda) = (1/m)\ \textstyle\sum_k\ (k\lambda - m)\ \Pi_{n-k}(\lambda),$$
>
> where the summation extends from $k=1$ to $\lfloor m/2 \rfloor$.

CHROMATIC POLYNOMIALS

BY

G. D. BIRKHOFF AND D. C. LEWIS

TABLE OF CONTENTS

The opening of Birkhoff and Lewis's paper on chromatic polynomials.

Chapter II then included a systematic list of the chromatic polynomials of all 111 cubic maps with 6 to 17 regions.

The next two chapters presented inequalities for the coefficients, with results that were related to the earlier contributions of Birkhoff and Whitney (see Chapters 2 and 4). Also featured here was a conjecture whose proof would imply the four color theorem. Noting that for a cubic map with $n-3$ regions, $\lambda(\lambda-1)(\lambda-2)$ must divide the chromatic polynomial $P(\lambda)$, the authors proposed that

$$(\lambda-3)^n << P(\lambda)/(\lambda(\lambda-1)(\lambda-2)) << (\lambda-2)^n, \text{ for } \lambda \geq 4,$$

where $f(\lambda) << g(\lambda)$ when the coefficients of the polynomial f are non-negative and do not exceed the corresponding coefficients of the polynomial g. Using their data from Chapter II, they verified the truth of their conjecture when $n \leq 8$, and also whenever $\lambda \geq 5$. Expressing the hope that it may "eventually turn out to be easier to establish", they continued:

> It is also hoped that the theory of the chromatic polynomials may be developed to the point where advanced analytic function theory may be profitably applied.

The final two chapters of their paper recalled Birkhoff's earlier use of Kempe-chain arguments to investigate the reducibility of configurations surrounded by rings with 4, 5, and 6 regions (see Chapter 2), and showed how to obtain the same results in a different way, asserting that:[3]

> Undoubtedly numerous other similar configurations can be proved to be reducible by the same methods, which are characteristic of the quantitative point of view . . . Thus the present work can to some extent be regarded as an attempt to bridge the gap between two previously separated points of view.

For years afterward, the Birkhoff–Lewis paper would be an authoritative resource for the study of chromatic polynomials, with many later authors citing it in their writings. However, it was not easy to read, as W. T. Tutte would later recall:[4]

> They do give a partial theory of these equations in their paper, but I confess that I was never able to read right through it and understand it clearly.

Arthur Bernhart

Arthur Frederick Bernhart was born in 1908. After studying at Olivet College in central Michigan, he transferred to the University of Michigan in Ann Arbor, where in 1934 he was awarded his doctoral degree in mathematical physics and quantum theory for a thesis on the mechanics of a top. He published papers on geometry and was an early authority on curves of pursuit. In 1943, he joined the mathematics faculty of the University of Oklahoma, where he remained until his retirement.

Captivated by map coloring problems, and inspired by Birkhoff and Franklin's detailed studies of reducible configurations, he wrote the significant paper "Six-rings in minimal five-color maps".[5] Here he confirmed Birkhoff's results on rings with four or five regions, before continuing Birkhoff's investigations into configurations surrounded by a ring with six regions. His detailed and exhaustive analysis showed that these configurations were of six types, which he called "solutions"; the first three of these consisted of rings with six regions that, respectively, surround a single hexagon, two pentagons sharing a common edge, and three pentagons sharing a common vertex.

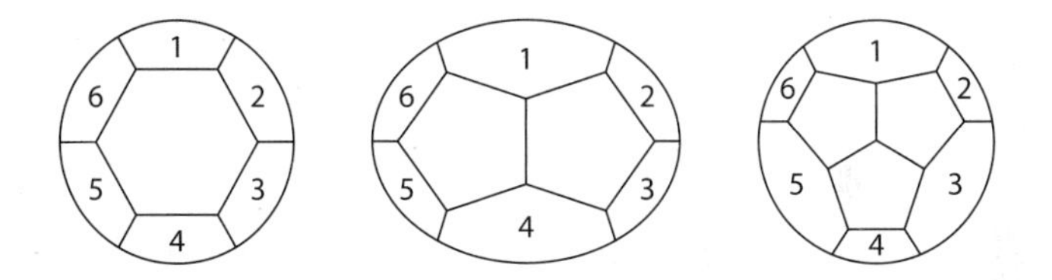

Three of Bernhart's "solutions" for rings with six regions.

In a second paper on reducibility, "Another reducible edge configuration",[6] Arthur Bernhart used a Kempe-chain argument to prove the reducibility, in a cubic map, of the configuration that consists of an edge common to two hexagons which border two pentagons.

On rare occasions, the map coloring disease has been hereditary. After attending the Universities of Oklahoma and Michigan, Arthur Bernhart's son, Frank, was awarded a doctoral degree from Kansas State University in 1974 for a thesis on map coloring. He then undertook postdoctoral work at the University of Waterloo in Ontario, before moving to teaching positions at a number of other colleges and universities. During this time, he discovered several significant results on graph coloring, and proved, for example, that a graph is 5-colorable if it becomes planar whenever any edge is removed.[7] He also gained an enviable reputation for his remarkable ability to uncover flaws in attempts on the four color problem, with one of his papers pointing out errors that had already appeared in two published "solutions" of the problem.[8]

By the middle of the 20th century, good progress had been made on solving the four color problem. But although those who followed in the footsteps of Birkhoff, Franklin, Errera, Reynolds, and Winn discovered many hundreds of reducible configurations, relatively little progress had been made on the construction of unavoidable sets. Indeed, apart from Kempe, Wernicke, and Franklin, only one further investigator seems to have contributed to this area by presenting several new unavoidable sets in 1940; this was Henri Lebesgue, best known for his theory of integration (the "Lebesgue integral") in the last paper that he ever wrote.[9] There was then a lull in activity, as significant new contributions to solving the four color problem were not to appear until the 1960s, leading to its eventual solution in 1976.

Isidore N. Kagno

Isidore Noah Kagno was born in 1908 and graduated in mathematics from Columbia University in 1930, receiving his master's degree in the following year for an essay "On the isomorphism between an algebra of symbols and the algebra of points on a line". He was awarded his doctoral degree from Columbia in 1939 for a thesis on topology, *Perfect Subdivisions of Surfaces*, which was also published.[10] Earlier, he had written a number of other papers on topological graph theory.[11] These included "The mapping of graphs on surfaces", in which he constructed graphs that cannot be embedded on the torus, the projective plane, and other surfaces; "The triangulation of surfaces and the Heawood color formula", where he investigated the Heawood conjecture; and the note (mentioned in Chapter 3) in which he showed that Tietze's formula for coloring maps on the non-orientable surface N_q holds when q is 3, 4, and 6.

Much of the aesthetic appeal of graph theory derives from attractive drawings of graphs with many symmetries, such as the Petersen graph (see Interlude A). The symmetries of a graph can be specified by its *automorphism group*, which consists of those mappings from the graph to itself that preserve the adjacency of vertices. For example, the complete graph K_n has $n!$ automorphisms, corresponding to the $n!$ permutations of the set of vertices, while the Petersen graph with ten vertices has $5! = 120$ automorphisms. We met groups of graphs briefly in the work of H. Roy Brahana (see Chapter 3), and in 1938 the German–Chilean mathematician Roberto Frucht proved that every abstract group is the automorphism group of some graph—indeed, of some cubic graph.[12] For this, he invoked the concept of a "Cayley color graph", which Arthur Cayley had introduced in 1878.[13]

In the 1940s, Kagno's interests turned to automorphism groups, and in 1946 he wrote his best-known paper, "Linear graphs of degree ≤ 6 and their groups",[14] in which he determined the automorphism groups of all twenty-two connected simple graphs with up to six vertices (but with none of degree 1 or 2). He followed this with the paper "Desargues' and Pappus' graphs and their groups",[15] in which he cited some work on "Toroidal and non-toroidal graphs" by Sister Mary Petronia Van Straten,[16] who had written the doctoral thesis *The Topology of the Configuration of Desargues and Pappus* for Notre Dame University in Indiana, under the supervision of Karl Menger. Born in Wisconsin, she later taught at Mount Mary College in Milwaukee, and became president of the Wisconsin Mathematics Council. In great demand for her lectures on mathematics and its teaching, she was awarded the Outstanding Educator of America Award and appeared in *Who's Who of American Women*.

Sister Mary Petronia Van Straten (1913–87).

Richard Otter

In Chapter 1, we saw how Arthur Cayley had investigated trees and their connections with certain chemical molecules. In the first of his papers, in 1857, his interest was in counting rooted trees with a given number of vertices or edges.[17] By removing the root, and examining the rooted trees that result from doing so, he showed that, if r_n is the number of rooted trees with n vertices, then their generating function

$$r(x) = r_1x + r_2x^2 + r_3x^3 + r_4x^4 + r_5x^5 + \cdots = x + x^2 + 2x^3 + 4x^4 + 9x^5 + \cdots$$

satisfies the recurrence relation

$$r(x) = x \times (1-x)^{-r_1} \times (1-x^2)^{-r_2} \times (1-x^3)^{-r_3} \times \cdots.$$

By expanding each of these binomial expressions as far as necessary, he was then able to calculate the numbers r_n, one at a time.

Counting unrooted trees is more difficult, but Cayley was able to obtain information about their number by building up each tree from its central vertex or vertices.[18] If t_n is the number of unrooted trees with n vertices, then their generating function is

$$t(x) = t_1x + t_2x^2 + t_3x^3 + t_4x^4 + t_5x^5 + \cdots = x + x^2 + x^3 + 2x^4 + 3x^5 + \cdots.$$

But is there any connection between these two generating functions, $r(x)$ and $t(x)$?

This question was answered by Richard Robert Otter. After graduating from Dartmouth College in 1941, he transferred to Indiana University in Bloomington where he completed a doctoral degree in organic chemistry in 1946. His interests then changed to mathematics, and he spent a short time at Princeton University. In 1948, he was appointed to the mathematics faculty at Notre Dame University, where he taught until his retirement in 1985.

While still at Princeton, Richard Otter wrote a paper, "The number of trees", in which he explored the properties of generating functions in general before turning to the enumeration of trees in particular.[19] Dismissing the approach of George Pólya (see Chapter 3) as "superfluous for the treatment of trees and rooted trees alone", he observed:

> in this paper purely combinatorial methods are employed for the development of relations between the generating functions. These methods enable one to study some general problems concerning the number of trees and of rooted trees and to find recursion formulas for counting these objects. Furthermore, the method used here for the counting of trees is new and interesting and considerably simpler than the methods used in the past.

Otter's main result was the following remarkable connection between the generating functions $r(x)$ and $t(x)$:

$$t(x) = r(x) - \tfrac{1}{2}(r^2(x) - r(x^2)).$$

Now known as *Otter's formula*, this result gives us the simplest way of calculating the tree-counting numbers t_n.

Otter's paper contains a footnote in which he "wishes to express his gratitude to Professor E. Artin for the suggestion of this problem and for his encouraging help towards its solution". Emil Artin was a celebrated algebraic number theorist. An academic refugee from Hamburg, he had arrived in the United States in 1937 and taught for a year at Notre Dame University, before moving to Indiana University in Bloomington for eight years. In 1946, he took up a position at Princeton University, where he stayed until 1958 before returning to Germany. Coincidentally these were the same institutions as those attended by Otter, but it is not recorded when and where they collaborated on the counting of trees.

Claude E. Shannon

In Interludes A and B, we featured two results on coloring the edges of maps and graphs, with adjacent edges (those that share a common vertex) always colored differently. The first of these was Tait's observation from 1880 that the four color conjecture is true if and only if the edges of every cubic planar graph can be colored with just three colors. The second was König's result from 1916 that the edges of any bipartite graph with maximum vertex degree Δ can be colored with Δ colors. More than thirty years were then to elapse before the next substantial result on edge-colorings was presented, by one of the most important scientific figures of the 20th century, as we now see.

Claude Elwood Shannon graduated from the University of Michigan in 1936, with bachelor's degrees in electrical engineering and mathematics. He then moved to MIT where, at the age of 21, he effectively founded

Claude E. Shannon (1916–2001).

digital circuit theory by showing how all problems of Boolean logic can be solved by electrical circuits, using open and closed switches for the symbols 0 (true) and 1 (false). He next worked with the computer pioneer Vannevar Bush on the construction of a differential analyzer, and at Bush's suggestion transferred to the Cold Spring Harbor Laboratory in New York to write his doctoral thesis on a mathematical foundation for Mendelian genetics. In 1941, after a year as a National Research Fellow at Princeton's Institute for Advanced Study, he transferred to Bell Telephone Laboratories for war work on codebreaking and the mathematical theory of cryptography. This quickly led to fundamental work on telecommunications and, in 1948, to a seminal paper on "A mathematical theory of communication", which earned him the description of "the father of information theory".

In 1949, Claude Shannon wrote a paper on the color coding of wires in electrical units such as relay panels, which began:[20]

> In these units there are a number of relays, switches, and other devices $A, B, \ldots, E$ to be interconnected. The connecting wires are first formed into a cable . . . and it is necessary, in order to distinguish the different wires, that all those coming out the cable at the same point be differently colored.

Assuming that at most m wires can emerge at any one point, Shannon proved that $\lfloor \frac{3}{2}m \rfloor$ colors are sufficient to color all the wires, and gave examples where this many colors are needed. Expressed in the language of graphs, his result asserts that the edges of any graph with multiple edges and maximum vertex-degree Δ can be colored with $\lfloor \frac{3}{2}\Delta \rfloor$ colors, with adjacent edges colored differently. Shannon's result on edge-colorings was not improved on until the mid-1960s, when discoveries by the Russian mathematician Vadim G. Vizing completely transformed the subject.[21]

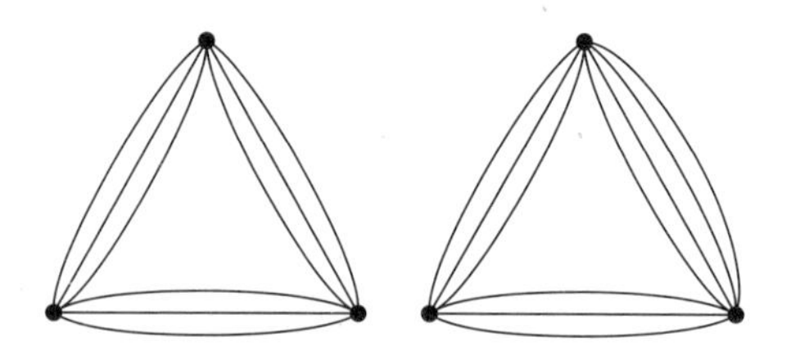

Two multigraphs for which Shannon's bound is achieved.

W. T. TUTTE

Bill Tutte is remembered for his very considerable contributions at Bletchley Park to the Allies' victory in World War II, and also for his groundbreaking work in graph theory and related topics. For a firsthand account of his fascination for the latter, see his autobiographical *Graph Theory as I Have Known It*.[22] Further information about his life and work can also be found in the biographies by his Waterloo friend and colleague Dan Younger and by Arthur Hobbs and James Oxley,[23] and in the two-volume *Selected Papers of W. T. Tutte*.[24]

William Thomas Tutte was born in Newmarket, near Cambridge, and enjoyed a successful high school career. While at school he developed an interest in mathematical puzzles, as he later recalled:[25]

> I was chiefly interested in the scientific subjects, including mathematics. One stimulus that seems to me supremely important came from outside the regular curriculum. In the school library I came upon a copy of Rouse Ball's Mathematical Recreations and Essays [Ball 1892]. There I found much information about graph theory and more general combinatorics. I read the basic theory of the Four Colour Problem and a discussion, without proofs, of Petersen's work on cubic graphs. This was my first encounter with the subject in which I was later to specialize.

In 1935, Bill Tutte entered Trinity College, Cambridge, to read natural sciences, and although his undergraduate studies were in chemistry,[26]

> I did have an interest in mathematical problems, strong enough to make me join the Trinity Mathematical Society. I often talked about such problems with three other members of the Society, students of Mathematics. They were Leonard Brooks, Cedric Smith and Arthur Stone. Each was destined to make his mark in Graph Theory.

The four of them became the closest of friends, and Tutte found that:[27]

> As time went on, I yielded more and more to the seductions of Mathematics.

In 1938, Tutte graduated with a first-class honors degree in chemistry and began postgraduate study in physical chemistry in Cambridge University's Cavendish Laboratory, working on spectroscopy and writing short notes for *Nature*.[28] But increasingly he came to realize that he would never make any headway as an experimental scientist and that his future should lie in mathematics. So, at the end of 1940, his tutor, Patrick Duff, arranged for Trinity College to transfer him from science to mathematics. But by then, World War II had already lasted for over a year, and Duff, recognizing Tutte's brilliance, arranged for him to be interviewed by the highly secret organization at Bletchley Park.

Bletchley Park housed the wartime headquarters of the British Government Code and Cypher School (now called GCHQ), where German airborne communications and those of the Italians and Japanese were monitored by Alan Turing and his associates. Bill Tutte joined the unit as a codebreaker in January 1941 and remained there for the duration of the war. In that year, the Germans began to replace their earlier Enigma cipher machines by a state-of-the-art Lorenz machine (codenamed "Tunny" by the British codebreakers) for sending intelligence messages from the German army's high command. With messages coming directly from Hitler, Rommel, and others, the greatest degree of secrecy was essential. The Tunny machine was necessarily highly complex, but in just a few months, the 24-year-old Tutte succeeded in unraveling its internal workings—an astonishing feat of cryptanalysis that is sometimes believed to have shortened the war by two years or more.

After the war was over, Tutte returned to academic life in Cambridge. Elected to a research fellowship in mathematics at Trinity College, he submitted no fewer than six articles for publication in his first year, on Hamiltonian graphs, factorization, the dichromate (or "Tutte polynomial"), and other topics. At the same time, he began to work on his doctorate, eventually producing a remarkable 417-page dissertation on *An*

W. T. Tutte at Bletchley Park and at the University of Waterloo.

Algebraic Theory of Graphs that appeared in 1948 and combined a wide range of ideas from algebra and combinatorics:[29]

> My thesis attempted to reduce Graph Theory to Linear Algebra. It showed that many graph-theoretical results could be generalized as algebraic theorems about structures I called 'chain-groups'. Essentially, I was discussing a theory of matrices in which elementary operations could be applied to rows but not columns.

His results represented the first major advances on matroids since their introduction in 1935 by Hassler Whitney (see Chapter 4) and are discussed in a survey of Tutte's contributions to matroid theory.[30] Over the coming years, the innovations in his thesis would give rise to a number of significant publications that put him at the forefront of this subject.

After receiving his doctoral degree from Cambridge University, Bill Tutte was invited by the well-known geometer H.S.M. (Donald) Coxeter to cross the Atlantic Ocean and join the University of Toronto in Canada, first as a lecturer and later as an associate professor. During his fourteen years in Toronto, Tutte became increasingly well known in mathematical circles for his wide-ranging achievements in combinatorics.

In 1958, Tutte was elected a Fellow of the Royal Society of Canada, and four years later he was invited by Ralph G. Stanton to join the mathematics faculty of the University of Waterloo in Ontario. Established in 1959, the university had been gradually enhancing its identity and standing within the academic world, and in 1967, it created what would become its highly regarded Department of Combinatorics and Optimization. Under the direction of Bill Tutte, its reputation soon attracted combinatorialists from all around the world. Meanwhile, in 1966, the *Journal of Combinatorial Theory* had been launched, with Tutte as its editor in chief. He retired in 1985.

A prolific writer of papers and books, even though his writing style made some of his publications difficult to understand, Bill Tutte published on a wide range of topics. We now look at a selection of these.

Squaring the Square

> Once upon a time there were four undergraduates of Trinity College, Cambridge, and they took as their hobby the study of perfect rectangles. By a perfect rectangle they meant a rectangle that is dissected into unequal squares . . . They hoped to find a perfect square, a perfect rectangle that was itself a square. A conjecture was going around that such a figure was impossible.

So recalled Bill Tutte[31]—and, as we mentioned earlier, the four undergraduates were Leonard Brooks, Cedric Smith, Arthur Stone, and Tutte himself. Stone had learned about the "squaring the square" problem from his tutor, who in turn had heard of it at a lecture in Cambridge given by Paul Erdős. The four friends were determined to settle it.[32]

In this problem, all squares are required to have integer-length sides. The problem was first solved for rectangles in 1925 by a Polish student named Zbigniew Moroń,[33] who divided a 33×32 rectangle into nine unequal squares—but no one seemed able to solve it for squares. While struggling with the intricacies of the problem, the "Gang of Four" (as they sometimes named themselves) soon discovered an unexpected connection between squared rectangles and electrical networks, with the horizontal levels of the various squares as vertices that were joined by downward arcs labeled with the sizes of the squares.

These networks (or "Smith diagrams") satisfy Kirchhoff's current laws, in that the sum of the numbers into any vertex (other than the top and bottom ones) is equal to the sum of the numbers out of it, and the sum of the numbers around each cycle is 0. Working with the networks proved to be simpler than searching for the squares directly, and the Gang of Four eventually managed to discover solutions, such as a dissection of a 61×69 rectangle into nine unequal squares. After much further effort, they came across what they believed to be the first squared square, with 26 unequal squares and side-length 608, but unfortunately for them, they were beaten by Roland Sprague of Berlin,[34] who had just

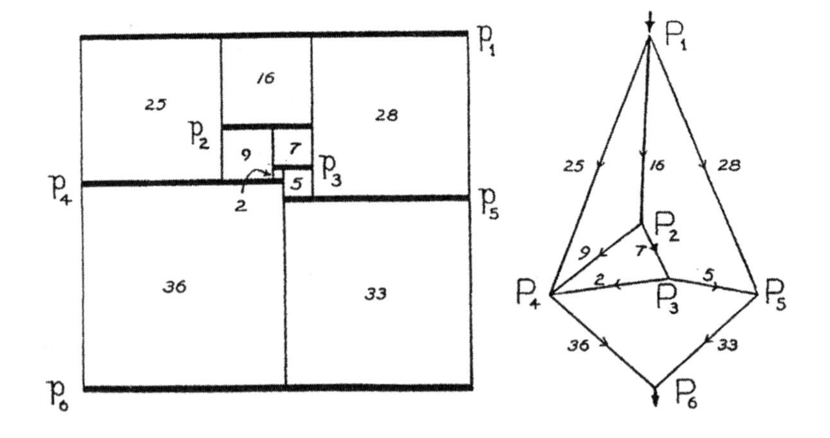

The Gang of Four's 61×69 squared rectangle and its associated network.

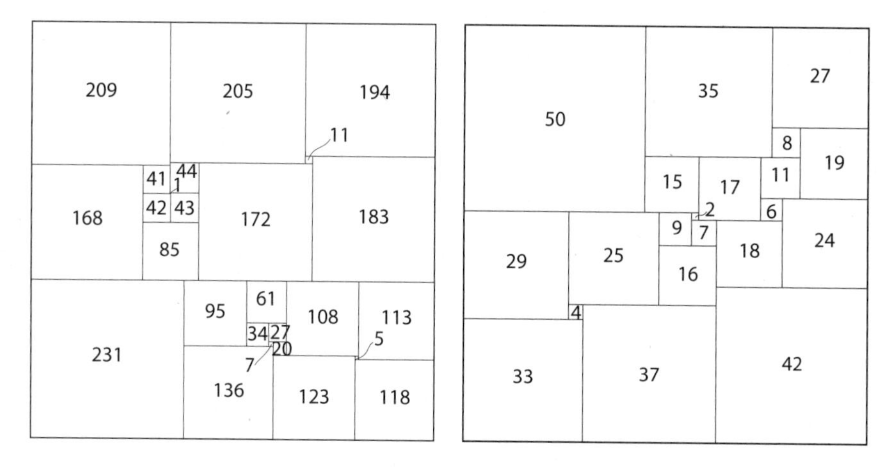

Squared squares by the Gang of Four and Arie Duijvestijn.

found one with 55 unequal squares and side-length 4205. It is now known that the smallest possible number of squares is 21, and an example with this number was found in 1978 by the Dutch computer scientist Arie Duijvestijn.[35]

The Gang of Four remained close friends for the rest of their lives. In 1941, Leonard Brooks proved "Brooks's theorem",[36] an important result on coloring the vertices of a graph. Cedric Smith, a statistician and geneticist, became the Weldon Professor of Biometry at University College, London. Arthur Stone obtained a doctoral degree at Princeton University before returning to England to teach at Trinity College, Cambridge, and Manchester University; a distinguished topologist, he migrated to the United States in 1961, where he taught at the University of Rochester in New York. Meanwhile, Bill Tutte continued to write on squaring the square and related puzzles for many years to come, including an article in 1950 that delved more deeply into the methods that underlie their construction.[37]

Hamiltonian Graphs

In Interlude A, we presented P. G. Tait's claim of 1884 that every cubic polyhedron has a Hamiltonian cycle, a conjecture whose truth would have led to a simple proof of the four color theorem. For over sixty years the issue remained unresolved, but while still at Bletchley Park, Bill Tutte constructed the following counter-example with 25 regions, 46 vertices, and 69 edges. It was published in 1946.[38]

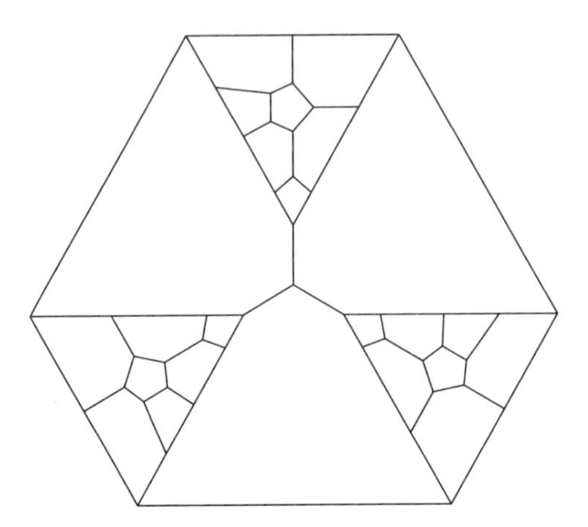

Tutte's example of a cubic polyhedron
with no Hamiltonian cycle.

A cubic polyhedron can be thought of as a 3-connected planar graph, and Tutte proved that if such a graph has any Hamiltonian cycles, then there must be at least three of them. In a later paper, building on ideas of Hassler Whitney, he proved further that if a planar graph is 4-connected, then it must always have Hamiltonian cycles.[39]

Factorization

In Interlude A, we defined a *1-factor* (or *perfect matching*) in a graph G to be a collection of non-adjacent edges meeting every vertex of G, and we presented Julius Petersen's sufficient condition for a cubic graph to have a 1-factor. We further defined an *r-factor* in G to be a regular subgraph of degree r that includes every vertex of G. In Interlude B, we presented Dénes König's result that every regular bipartite graph has a 1-factor.

In 1947, while working on his doctoral thesis, Bill Tutte significantly advanced the subject by presenting the following condition for a connected graph to have a 1-factor:[40]

> A connected graph G has a 1-factor if and only if, for each proper subset S of vertices, the number of odd components of $G-S$ is at most $|S|$.

Here, an "odd component" is a connected component with an odd number of vertices, and $G-S$ is the graph obtained from G by removing the vertices in S and their incident edges.

Unfortunately, Tutte's proof was somewhat clumsy, involving Pfaffian determinants, but soon after, Tibor Gallai and F. G. Maunsell presented simpler proofs that avoided them.[41] In 1952, Tutte returned to the topic and found a condition for a given graph to have an r-factor for any number r.[42] He realized that matters had become very complicated, and in a follow-up paper, he proved the surprising fact that the result for r-factors, rather than generalizing his result for 1-factors, could also be deduced from it.[43] To do so, he constructed a new graph G' from the original graph G, and proved that G has an r-factor if and only if G' has a 1-factor. The subject of factorization had been transformed in just five years.

Matroids

In Chapter 4, we saw how Hassler Whitney had introduced the concept of a matroid. This generalized ideas from both linear algebra and graph theory, with the properties of independence and bases arising from the former, and those of cycles arising from the latter. Although Whitney's 1935 paper had caused some limited interest, there was then little activity for over twenty years.

The situation changed dramatically in the late 1950s, with a succession of remarkable papers in which Bill Tutte built on results from his doctoral thesis.[44] These relaunched the subject and stimulated a flurry of activity, especially after links were discovered between matroids and matchings. Matroid theory is now part of mainstream combinatorial mathematics.

Whitney had concluded his paper with a section on matroids that arise from linear algebra. We say that a matroid M is *representable* over a field if it can arise from a set of vectors in some vector space over that field. If M is representable over every field, then it is called a *regular matroid*, and if it is representable over the field of two elements (0, 1) (modulo 2), it is a *binary matroid*. It can be proved that the dual matroid of a regular matroid is regular and of a binary matroid is binary.

Of particular importance is the *Fano matroid* on the set $E = (1, 2, 3, \ldots, 7)$, whose bases are all triples of numbers in E, except for the triples

$$(1, 2, 4), (1, 3, 5), (1, 6, 7), (2, 3, 6), (2, 5, 7), (3, 4, 7), (4, 5, 6).$$

These triples correspond to the lines of the finite projective plane of order 2, named the *Fano plane* after the Italian geometer, Gino Fano, who introduced it in 1892.[45] Whitney showed that this matroid is not representable over the field of real numbers, but that it is representable over the field of two elements (0, 1). It is therefore a binary matroid that is not regular.

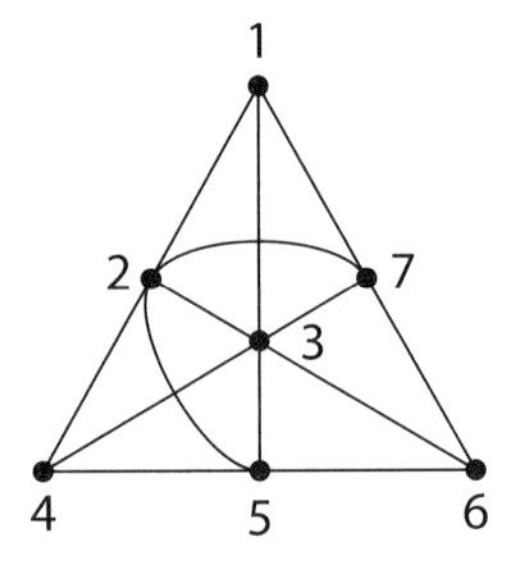

The Fano plane.

Before proceeding, we recall Kuratowski's theorem (from Interlude B) that a graph is planar if and only if it has no subgraph that is homeomorphic to the complete graph K_5 or the complete bipartite graph $K_{3,3}$. An alternative version was obtained by K. Wagner, and later by Frank Harary and Bill Tutte, based on the operations of deleting and contracting the edges of a graph.[46] If $e=vw$ is an edge of a graph G, we can obtain two more graphs from G:

deletion: $G-e$ is the graph we obtain by deleting e from G, but not the vertices v and w;

contraction: G/e is the graph we obtain by contracting e—that is, deleting e, and identifying v and w, so that all edges that were formerly incident to either are incident to the new vertex.

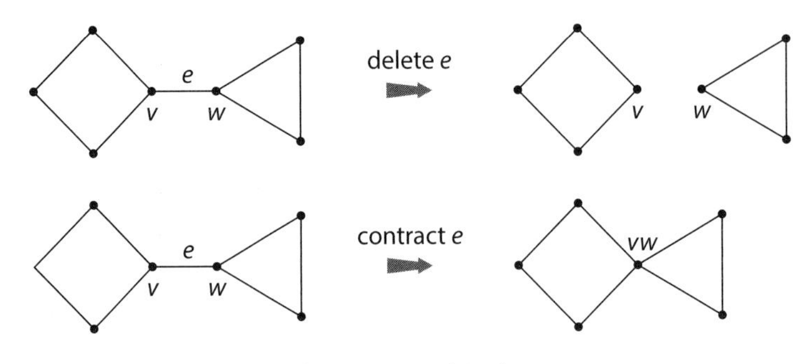

Deleting and contracting the edge e.

The alternative version of Kuratowski's theorem is then as follows:

Kuratowski's theorem: A connected graph is planar if and only if it cannot be reduced to K_5 or $K_{3,3}$ by a succession of edge deletions and contractions.

There are corresponding operations for matroids, and a *minor* of a matroid M is any matroid that can be obtained from M by a succession of such deletions and contractions.

Bill Tutte's great achievement was to find Kuratowski-like characterizations of the matroids that arise from graphs, as we now see.

W. T. Tutte: *Matroids and graphs* (1959)

We recall from Chapter 4 that two matroids are naturally associated with a connected graph G: its cycle matroid $M(G)$, whose circuits are the cycles of G, and its cutset matroid $M^*(G)$, whose circuits are the (minimal) cutsets of G. These matroids are duals of each other.

We say that a matroid is *graphic* if it is the cycle matroid of some graph, and *cographic* if it is the cutset matroid of some graph. A matroid that is both graphic and cographic is a *planar matroid*: these are the graphic matroids that arise from planar graphs. It can be shown that the cycle and cutset matroids of the two Kuratowski graphs K_5 and $K_{3,3}$ are neither graphic nor cographic.

We have seen that a binary matroid is one that is representable over the field of two elements (0, 1) (modulo 2). If G is a graph, then its cycle matroid $M(G)$ is a binary matroid. To see this, we associate with each edge of G the corresponding row of its incidence matrix, regarded as a vector with components 0 and 1. Then, if a set of edges forms a cycle in G, the sum (modulo 2) of the corresponding vectors is 0. It follows that we may have to restrict our attention to binary matroids when attempting to extend properties of graphs to matroids.

When is a given matroid graphic? We have seen that it must be a binary matroid (in fact, it is regular), and also that it cannot contain as a minor the Fano matroid F or its dual F^*, or either of the dual matroids $M^*(K_5)$ and $M^*(K_{3,3})$. Tutte proved that these necessary conditions are sufficient:

> A matroid M is graphic if and only if it is binary and has no minor that is isomorphic to $M^*(K_5)$, $M^*(K_{3,3})$, F, or F^*.

Applying this result to the dual matroid M^*, and recalling that the dual of a binary matroid is binary, we deduce that:

> A matroid M is cographic if and only if it is binary and has no minor that is isomorphic to $M(K_5)$, $M(K_{3,3})$, F, or F^*.

In an earlier paper, Tutte had also proved the following very deep result:

> A binary matroid is regular if and only if it has no minor that is isomorphic to the Fano matroid F or its dual F^*.

Combining these results, we obtain Tutte's matroid analog of Kuratowski's theorem:

> A matroid is planar if and only if it is regular and has no minor that is isomorphic to $M(K_5)$, $M(K_{3,3})$, or their duals.

The Tutte Polynomial

While examining the networks associated with the "squaring a square" problem, the four Trinity College students discovered that if $C(G)$ is the number of spanning trees in a connected graph G, then for any edge e, $C(G)$ satisfies the recursion formula

$$C(G) = C(G-e) + C(G/e),$$

where $G-e$ and G/e are obtained by deletion and contraction, as explained earlier.

In a similar way, if $P(G, \lambda)$ is the chromatic polynomial of a connected graph G, and if $e = vw$ is an edge of G, then, by counting the number of colorings in which the vertices v and w have the same, or different, colors, we have

$$P(G, \lambda) = P(G-e, \lambda) - P(G/e, \lambda),$$

or, if $Q(G, \lambda) = a(G)\, P(G, \lambda)$, where $a(G) = 1$ or -1, according to whether the number of vertices of G is even or odd, then

$$Q(G, \lambda) = Q(G-e, \lambda) + Q(G/e, \lambda).$$

This provides an unexpected link between squaring the square and chromatic polynomials, and also with another polynomial called the "flow polynomial", which is a sort of dual of the chromatic polynomial.

Tutte became very interested in recursion formulas of the form

$$W(G) = W(G-e) + W(G/e),$$

where W is a graph parameter, and in his thesis and in a paper written in 1947,[47] he investigated their properties. As he recalled later:[48]

> Playing with my W-functions I obtained a two-variable polynomial from which either the chromatic polynomial or the flow-polynomial could be obtained by setting one of the variables equal to zero, and adjusting signs. With minor simplifications this became a function $T(G; x, y)$ . . . In my papers I called this function the dichromate, but it is now generally known as the Tutte polynomial. This may be unfair to Hassler Whitney who knew and used analogous coefficients without bothering to affix them to two variables.

These "analogous coefficients" were Whitney's numbers m_{ij}, which recorded the number of subgraphs with rank i and nullity j (see Chapter 4). In his calculations of chromatic polynomials, Whitney had used an observation of R. M. Foster's that the numbers m_{ij} satisfy the recursion formula

$$m_{ij}(G) = m_{ij}(G - e) + m_{i-1,j}(G/e),$$

for any graph G and edge e. Recognizing this as a W-function led Tutte to define the *Whitney rank-generating function* of a graph G with edge-set E, as

$$R_G(x, y) = \sum_{A \subseteq E} x^{r(E) - r(A)} y^{|A| - r(A)},$$

where $r(A)$ is the rank of A. Here, the coefficient of each term $x^i y^j$ counts the number of subgraphs with rank i and j edges. The *Tutte polynomial* is then given by

$$T_G(x, y) = R_G(x - 1, y - 1) = \sum_{A \subseteq E} (x - 1)^{r(E) - r(A)} (y - 1)^{|A| - r(A)}.$$

It can also be defined recursively by defining

$T_G(x, y) = 1$, if G has no edges,
$T_G(x, y) = x\, T_{G-e}(x, y)$, if $G - e$ is disconnected,
$T_G(x, y) = y\, T_{G/e}(x, y)$, if e is a loop,

and successively using the recursion relation

$$T_G(x, y) = T_{G-e}(x, y) + T_{G/e}(x, y).$$

The Tutte polynomial has the following properties for a connected graph G:

$T_G(1, 1)$ is the number of spanning trees in G.
$T_G(1, 2)$ is the number of spanning subgraphs.

$T_G(2, 1)$ is the number of forests (cycle-free subgraphs) in G.

$T_G(2, 2) = 2^m$, where m is the number of edges of G.

The chromatic polynomial of G is $\lambda\,(-1)^{r(E)}\,T_G(1-\lambda, 0)$.

If G is a planar graph, and G^* is a dual graph of G, then $T_G(x, y) = T_{G^*}(y, x)$.

Because every matroid can be defined in terms of a rank function, Tutte realized that matroids form the most natural setting for the above ideas, and showed how this could be done—noting, for example, that

> Every matroid M has a dual M^*, and $T_M(x, y) = T_{M^*}(y, x)$, for all x and y.

Tutte polynomials have also found significant applications in areas as varied as knot theory, coding theory, and statistical mechanics.

Chromatic Polynomials

We have seen how George Birkhoff and Hassler Whitney attempted to settle the four color problem for planar graphs or maps by proving that $P(4) > 0$, where $P(\lambda)$ is the chromatic polynomial. Because few planar graphs are 3-colorable, the values $P(1)$, $P(2)$, and $P(3)$ are usually 0, and in 1968, Bill Tutte sought to increase his understanding of chromatic polynomials by investigating other zeros that they may have. Concentrating mainly on large cubic planar graphs and triangulations, he amassed hundreds of chromatic polynomials, mainly from the doctoral thesis of Ruth A. Bari, written at Johns Hopkins University under the supervision of Daniel C. Lewis, and from the collection of Dick Wick Hall of the State University of New York at Binghamton.

With the help of his Waterloo colleague, Gerald Berman, Tutte obtained all the zeros of these polynomials from the university's IBM 700 computer. Analyzing this wealth of data,[49] they then made some surprising discoveries that involved the golden ratio,

$$\tau = \tfrac{1}{2}(1 + \sqrt{5}) = 1.618034\ldots, \text{ where } \tau^2 = \tau + 1.$$

In particular, they noted that, for large planar graphs, most chromatic polynomials seem to have zeros that lie very close to τ, and to the negative number $\tau - 2 = \tfrac{1}{2}(-3 + \sqrt{5}) = -0.381966\ldots$.

Tutte then demonstrated[50] that chromatic polynomials $P(\lambda)$ of large graphs also tend to have zeros that lie very close to the value

$$\lambda = \tau^2 = \tau + 1 = \tfrac{1}{2}(3 + \sqrt{5}) = 2.618034\ldots,$$

by proving that $P(\tau+1)$ is extremely small—in fact, for a triangulation with n vertices,

$$0 < |P(\tau+1)| \leq \tau^{5-n}.$$

Moreover, for

$$\lambda = \tau + 2 = \tau\sqrt{5} = \tfrac{1}{2}(5 + \sqrt{5}) = 3.618034\ldots,$$

Tutte proved what he called his *golden identity*:[51]

$$P(\tau+2) = (\tau+2)\, \tau^{3n-10} P^2(\tau+1).$$

It follows that $P(\lambda)$ is positive when $\lambda = 3.618034\ldots$, and also when $\lambda = 4$ (assuming the four color theorem). One might wonder whether $P(\lambda)$ also remains positive between these values, but this is not the case.

In 1975, Sami Beraha wrote a doctoral thesis for Johns Hopkins University, observing that the zeros of chromatic polynomials of large planar graphs tend to congregate around the numbers

$$B_n = 2 + 2 \cos 2\pi/n, \text{ for } n = 2, 3, \ldots,$$

and these numbers are now known as "Beraha numbers". For small values of n, they are

n:	2	3	4	5	6	7	8	9	10
B_n:	0	1	2	2.618 ...	3	3.247 ...	3.414 ...	3.532 ...	3.618 ...

Tutte was interested in these numbers, but it remains unknown as to why they seem to play such an important role in the theory of chromatic polynomials. Meanwhile, his interest in graph colorings continued for many years, and in Chapter 6 we shall see how he became involved with the four color theorem.

* * * * *

Bill Tutte has rightly been acknowledged as the postwar leader of combinatorial thinking, becoming North America's natural successor to Oswald Veblen, George Birkhoff, and Hassler Whitney. In the Foreword to Tutte's *Selected Papers*, Professor Ralph Stanton summed up his friend's achievements as follows:[52]

> I want to end this rather digressive foreword on a personal note, just to stress to readers the real significance of Bill Tutte's work. Not too many people are privileged to practically create a subject, but there have been several this century. Albert Einstein created Relativity . . . Similarly, modern Statistics owes its existence to Sir Ronald Fisher's exceptionally brilliant and creative work. And I think that Bill Tutte's place in Graph theory is exactly like that of Einstein in relativity and that of Fisher in Statistics. He has been both a great creative artist and a great developer.

ALGORITHMS

An *algorithm* (named after the 9th-century Persian mathematician, al-Khwārizmī) is a finite step-by-step process for solving a mathematical problem. We can think of it as a recipe, where we are given some input (ingredients), apply the algorithm (the steps of the recipe), and produce the required output (a perfect dish). The earliest graph theory algorithms, dating from the late 19th century, seem to be those of M. Fleury for producing an Eulerian trail, and the maze-tracing algorithms of Gaston Tarry and C. P. Trémaux.[53] Methods for tracing mazes were also discussed in the books by André Sainte-Laguë and Dénes König, mentioned in Interlude B, but few other algorithms appeared in graph theory publications in the early 20th century.

The situation changed markedly in the 1940s and 1950s, when systematic solutions were needed for a range of practical problems, arising first from World War II requirements and then as a consequence of the rapidly expanding computer industry. In this section, we outline the development of specific algorithms for investigating problems on matching, assignment, and transportation, finding flows in capacitated networks and minimum spanning trees, and solving path problems. Further information can be found in the excellent historical survey by Alexander Schrijver.[54]

Matching and Assignment

In Interlude B, we presented Philip Hall's "marriage" theorem:

> *Hall's theorem*: Suppose that each of a collection of boys is acquainted with a collection of girls. Then each boy can marry one of

his acquaintances if and only if, for each number k, every set of k boys is collectively acquainted with at least k girls.

This is a theorem on *matchings*, and can be illustrated by a bipartite graph in which one set of vertices corresponds to the boys and the other set corresponds to the girls, with edges joining the boys to the girls with whom they are acquainted. Similar graphs can be used in *assignment problems*, where a number of candidates apply for various jobs, and the bipartite graph has vertices for the applicants and the jobs, with edges joining the applicants to jobs for which they are qualified. The aim is to assign the greatest possible number of applicants to suitable jobs.

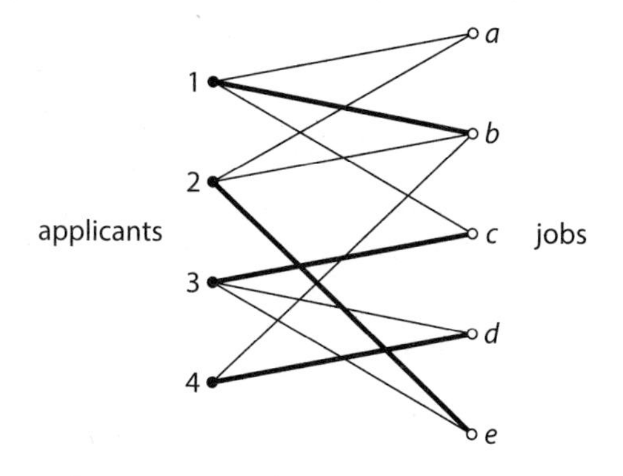

A maximum matching in a bipartite graph.

In such assignment problems, there may be several ways of matching the applicants to suitable jobs, and we may wish to determine which one is the "best". Because each applicant may be better suited to some jobs than others, we associate a "cost" or "weight" with each edge of the bipartite graph—the smaller the cost, the greater is the applicant's suitability for that job. The aim is then to find the assignment with the least total cost.

In Interlude B, we also presented König's minimax theorem on the largest size of a matching in a bipartite graph:[55]

König's theorem: In any bipartite graph, the maximum size of a matching is equal to the minimum number of vertices that collectively meet all the edges.

This theorem can be used to find a maximum matching in any given bipartite graph, and it was then generalized to weighted bipartite graphs by Jenő Egerváry.[56] Although Egerváry did not specifically mention an *algorithm*, his proof easily gives rise to one (though not an efficient one) for finding a maximum matching in a weighted bipartite graph.

Following the papers of König and Egerváry, assignment problems were ignored for several years. But in 1955, Harold W. Kuhn of Bryn Mawr College (near Philadelphia), and later of Princeton University, reexamined their work. After presenting the assignment problem, he added:[57]

> It is shown that ideas latent in the work of two Hungarian mathematicians may be exploited to yield a new method of solving this problem.

Using their ideas, Kuhn was able to develop an efficient algorithm for solving assignment problems, and named it the *Hungarian method*, a name that remains to this day. Kuhn's pioneering work was not expressed in the language of graph theory, but such a presentation can be found in a historical account by András Frank.[58] As Kuhn later recalled:[59]

> Using Egerváry's reduction and König's maximum matching algorithm, in the fall of 1953 I solved several 12 by 12 assignment problems (with 3-digit integers as data) by hand. Each of these examples took under two hours to solve and I was convinced that the combined algorithm was 'good'. This must have been one of the last times when pencil and paper could beat the largest and fastest electronic computer in the world.

Shortly after developing the Hungarian method, Kuhn moved to Princeton University. There, he had a successful career, developed many ideas that are named after him, received a major award for his contributions to game theory, and enjoyed a long and fruitful association with the Nobel Prize winner John Nash.

Transportation and Linear Programming

It is now known that the Hungarian algorithm had already been solved around 1836 by the German mathematician Carl Jacobi and published posthumously. The origins of the *transportation problem* and linear programming can similarly be traced back to Europeans. In 1794, the French geometer Gaspard Monge investigated a problem of minimizing the cost of transporting quantities of earth from one area to another, and in the 1820s, Joseph Fourier was experimenting with the solution of linear inequalities. Both of these topics resurfaced in America in the 1940s.

In 1941, the Harvard graduate and MIT mathematician Frank L. Hitchcock was analyzing the optimal distribution of various commodities through a network at minimum cost, and discovered a method for solving the following general transportation problem:[60]

> When several factories supply a product to a number of cities we desire the least costly manner of distribution. Due to freight rates and other matters the cost of [sending] a ton of product to a particular city will vary according to which factory supplies it, and will also vary from city to city.

This is similar to the assignment problem for bipartite graphs, in which each edge still has a given cost, but there are now additional supply requirements out of each factory vertex, and demand requirements into each city vertex.

Meanwhile, the Russians A. N. Tolstoĭ and Leonid Kantorovich had solved a range of transportation problems for the Soviet railway network, and Tjalling Koopmans, a Dutch mathematician and economist who had moved to the United States in 1940, was involved with problems of merchant shipping during World War II. Such problems led to maximizing or minimizing a function of variables that are subject to given equations or inequalities, and as the 1940s progressed, attempts to formulate them in this way led to a more efficient method of solution. Koopmans would later call this process *linear programming*. Many years later, in 1975, Kantorovich and Koopmans were jointly awarded a Nobel Prize for their achievements.

At this stage, George B. Dantzig, the "father of linear programming", entered the scene. After earning his first degrees from the Universities of Maryland and Michigan, he worked on an urban study at the US Bureau of Labor Statistics in Washington, DC. Two years later, he moved to San Francisco to work on a doctorate in statistics at Berkeley, where he amazed everyone by solving two famous unsolved problems in statistics: he believed them to be homework problems which had been set for a course that he was studying. With the outbreak of World War II, he moved back to Washington and became head of the Combat Analysis Branch of the US Army Air Forces Statistical Control, carrying out important work for which he received an Exceptional Civilian Service Award medal. After the war, he returned briefly to California to receive his doctoral degree, but was soon back in Washington as mathematical advisor to the Defense Department in the Pentagon.

It was around this time, in 1947, that Dantzig made his best-known contribution to mathematics. Building on the methods of Hitchcock, Kantorovich, and Koopmans, and on his own work with the US Air Force, he invented the *simplex method* for solving linear programming problems, and demonstrated its power by solving an assignment problem with 70 applicants and 70 jobs.[61] A meeting between Dantzig and John von Neumann, the mathematician and computer pioneer, led to the introduction of *duality* in linear programming, an important idea that had already been implicit many years earlier in the writings of Fourier.

In 1952, Dantzig moved to the RAND Corporation in California, where he worked on the computer implementation of linear programming. In 1960, he was back at Berkeley, where he set up the Operations Research Center, and then transferred to Stanford University, where he spent the rest of his career. Further information about his life and works can be found in his classic book *Linear Programming and Extensions* and in his historical reminiscences.[62]

Flows in Networks

In 1954, Ted Harris, a mathematician at the RAND Corporation, posed the following problem:

> Consider a rail network connecting two cities by way of a number of intermediate cities. If each link of the network has a number assigned to it, representing its capacity, find a maximal flow from one city to the other.

Problems of this kind arise whenever one wishes to maximize the amount of a commodity that is transmitted from a factory to a market along various channels.

Harris's problem arose in connection with a secret report that he and his colleague Frank Ross had been writing for the US Air Force, on the flow of rail traffic in the Soviet Union and Eastern Europe.[63] They split the railway network into a large number of arcs, each bearing a limited amount of traffic, and the aim was to find how much traffic could be carried by the whole network without any part of it becoming overloaded. Their initial approach was to send as many trains as possible and to deal with any bottlenecks as they arose. Eventually, they calculated a maximum flow of 163,000 tons from the Soviet Union to Eastern Europe, and located a bottleneck of the same total capacity through which it needed to flow.

Ted Harris mentioned the network flow problem to his colleagues Lester Randolph Ford Jr. and Delbert Ray Fulkerson. Ford had graduated from the University of Chicago and received his doctorate from the University of Illinois at Urbana–Champaign, while Fulkerson's doctoral degree had been awarded by the University of Wisconsin in Madison. By 1954, both were working for the RAND Corporation.

Lester Ford and Ray Fulkerson soon became interested in network flow problems. They realized that all such problems could be solved by the simplex method of linear programming, but sought a more direct

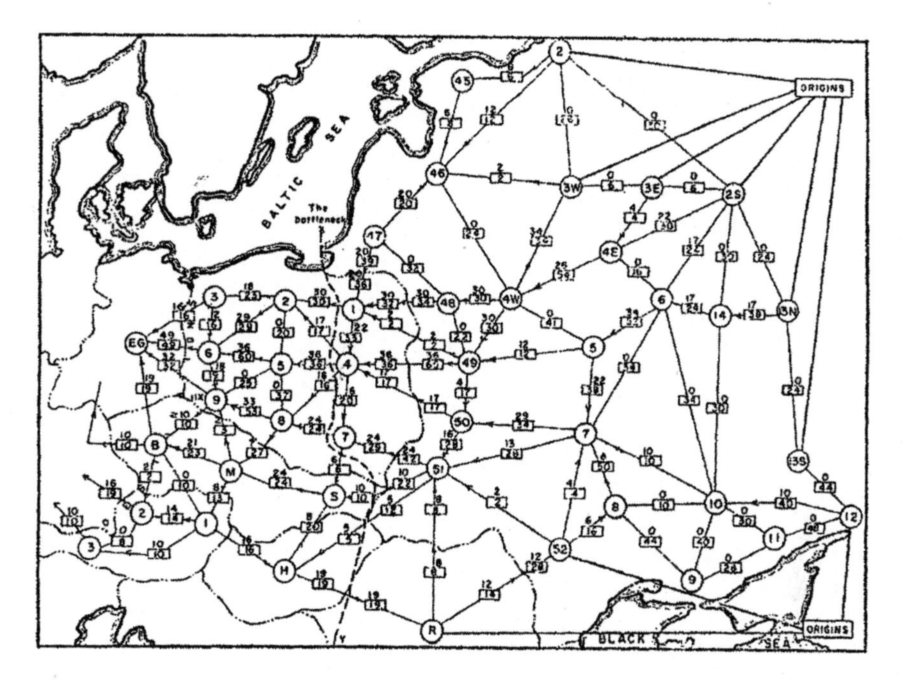

The railway network of Ted Harris and Frank Ross.

method. The results were a fundamental minimax theorem and an algorithm for finding a maximum flow. These appeared in a couple of RAND technical reports in 1954 and 1955, and were subsequently published in the *Canadian Journal of Mathematics*.[64]

L. R. Ford Jr. and D. R. Fulkerson:
***Maximal flow through a network* (1956)**

We can think of a *capacitated network* as a directed graph in which two vertices are designated as the *start vertex S* and the *terminal vertex T*, and where each arc has been assigned a positive number, called its *capacity*. A *flow* in the network is an allocation of a non-negative number to each edge, for which

the flow along each arc does not exceed the capacity of the arc;
the total flow into any vertex (other than S and T) equals the total flow out of it.

An arc is *saturated* if the flow along it equals its capacity, and the *value of the flow* is the total flow out of S, which equals the total flow into *T*. The aim is to find a *maximum flow*—that is, a flow from *S* into *T* with the largest possible value.

As an example, the following is a network flow in which the first number alongside each arc is the flow along it, and the second number is its capacity. Here, the arcs *CD* and *CT* are both saturated, and the value of the flow is 5.

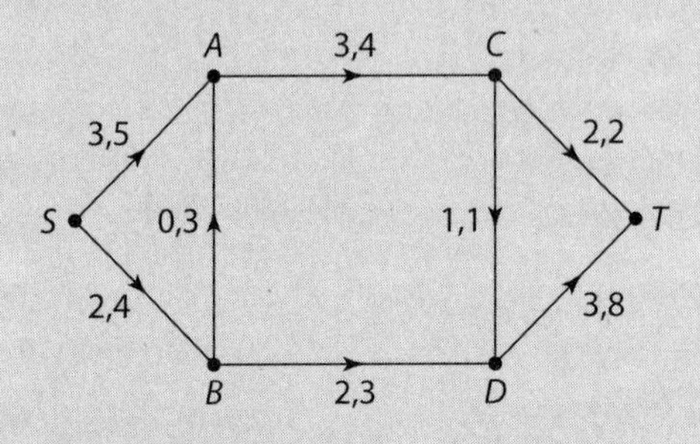

Ford and Fulkerson's main achievement in this paper was their "max-flow min-cut theorem". A *cut* in a network is a minimal set of arcs whose deletion splits the network into two parts, one containing *S* and the other containing *T*. The *capacity of the cut* is the sum of the capacities of those arcs in the cut that are directed from the part containing *S* to the part containing *T*; for example, in the above network, (*AC*, *BD*) is a cut with capacity 7, and (*BD*, *CD*, *CT*) is a cut with capacity 6. A *minimum cut* is a cut with the smallest possible capacity. It is easy to see that

> the value of any flow cannot exceed the value of any cut,

and it follows that

> the value of a maximum flow cannot exceed the capacity of a minimum cut.

The max-flow min-cut theorem states that these two numbers are always equal:

> *Max-flow min-cut theorem*: In a capacitated network, the value of a maximum flow from the start to the terminal is equal to the capacity of a minimum cut.

The following figure shows this network with a maximum flow of value 6 and a minimum cut of capacity 6; we note that minimum cuts always consist of saturated arcs.

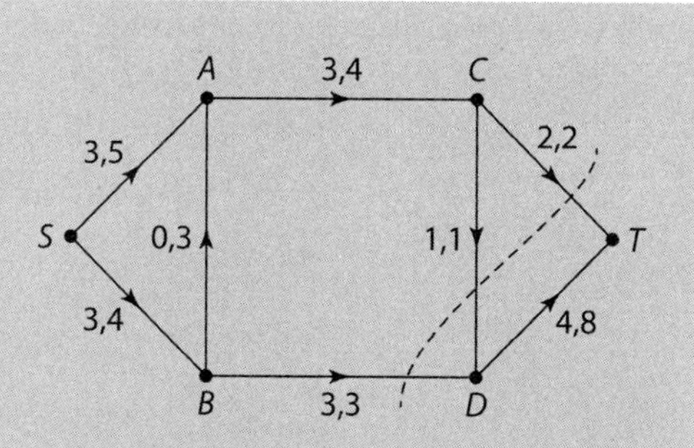

Ford and Fulkerson's second paper included an algorithm for finding a maximum flow. The aim is to accumulate *flow-augmenting paths*—that is, paths from S to T along which the flow can be increased. For example, starting with no flow in the above network, we can send

a flow of 2 along the path $S–A–C–T$,
a flow of 3 along the path $S–B–D–T$,
and a further flow of 1 along the path $S–A–C–D–T$.

This saturates the arcs BD, CD, and CT, and gives the maximum flow of value 6.

Ford and Fulkerson's methods can easily be adapted to deal with the following modifications to the network:

if the network contains any undirected edges, we replace each such edge by two arcs, one in each direction, and proceed as before;

if the network has several start vertices and/or terminal vertices, we add a new start vertex joined to all the start vertices, and a new terminal vertex to which all the terminal vertices are joined, and proceed as before;

if there are also capacity restrictions on the vertices (limiting the flow through a city, for example), we replace each such vertex V by two linked vertices, V_1 and V_2, with all incoming arcs to V directed toward V_1, and all outgoing arcs from V emerging from V_2.

We can also apply the max-flow min-cut theorem to the following situations:

to deduce Menger's theorem (see Interlude B), take the capacity of every edge to be 1;

to deduce Hall's theorem, add a new start vertex joined to all the boys, and a new terminal vertex joined to all the girls, and take the capacity of every edge to be 1;

to answer assignment or transportation problems, add a new start vertex joined to all the applicants (or factories), and a new terminal vertex joined to all the jobs (or cities).

The contributions of Ford and Fulkerson marked an important step in the development of combinatorial optimization, and the ensuing years witnessed other variations on the max-flow min-cut theorem, including a different proof by P. Elias, A. Feinstein, and C. E. Shannon.[65] In 1962, Ford and Fulkerson published their book *Flows in Networks*,[66] which quickly became the standard work on the subject.

Minimum Spanning Trees

In the mid-1950s, two American mathematicians, Joseph Kruskal Jr. and Robert Prim, presented algorithms for solving what is sometimes called the "minimum connector problem":

> *Minimum connector problem*: A railway network is to be built connecting several towns. If the distances between all pairs of towns are known, how can we design the network so that the total amount of railway track is as small as possible?

If we construct a weighted graph with vertices representing the towns, and weighted edges representing the portions of track connecting them, then we seek a set of edges with minimum total length that connects all the vertices. Because these edges can include no cycle, they must form a spanning tree. But for n towns, there are n^{n-2} spanning trees (by a result of Arthur Cayley), so how do we find the shortest one?

Joseph B. Kruskal Jr. graduated from the University of Chicago and transferred to Princeton University, where he was awarded his doctoral degree for a thesis on partially ordered sets in which he solved a problem put forth by Paul Erdős. For a period he worked for the Office of Naval Research at George Washington University and at the University of Wisconsin, before moving in 1959 to Bell Laboratories for the rest of his career.

It was while he was at Princeton that Kruskal wrote his paper on the *minimum spanning tree problem*, which he described as follows:[67]

> If a (finite) connected graph has a positive real number attached to each edge (the length of the edge), and if these lengths are all distinct, then among the spanning trees of the graph there is only one, the sum of whose edges is a minimum; that is, the shortest spanning tree is unique.

The method that Kruskal gave for finding the minimum spanning tree was as follows:

> CONSTRUCTION A. Perform the following step as many times as possible: Among the edges of *G* not yet chosen, choose the shortest edge which does not form any loops [cycles] with those edges already chosen. Clearly the set of edges eventually chosen must form a spanning tree of *G*, and in fact it forms a shortest spanning tree.

The mathematician and computer scientist Jack Edmonds later called this the *greedy algorithm*, because at each stage we make the "greediest" choice of an edge. He also noted that matroids form the natural setting for such algorithms.[68]

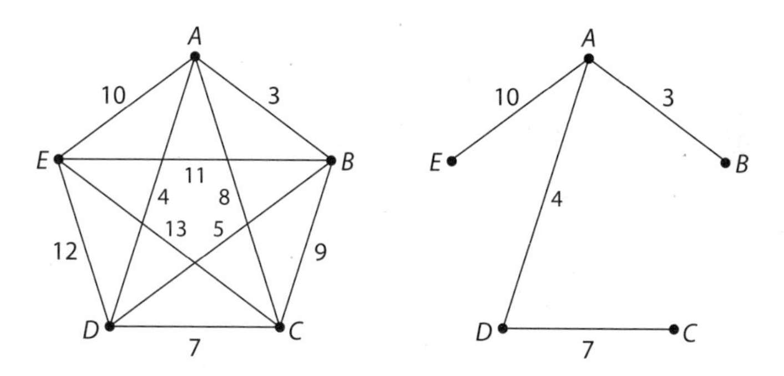

A minimum connector problem and its solution.

We illustrate the greedy algorithm with a simple example with five towns.

We first choose the edges *AB* (length 3) and *AD* (length 4).

We cannot then choose *BD* (length 5), because this would create the cycle *ABDA*, so instead we choose *CD* (length 7).

We cannot then choose *AC* (length 8), or *BC* (length 9), because either of these would create a cycle, and so instead we choose *AE* (length 10).

This completes the spanning tree, with minimum total length $3+4+7+10=24$.

It can be shown that if the edge-lengths are all different, as specified by Kruskal, then there is just one minimum connector. But if some edges are of equal length, then there may be several solutions, but they will all have the same total minimum length.

Kruskal continued with two more constructions for finding minimum spanning trees. In the first of these, we always choose the shortest edge between a town that we have already visited and a new one; for example, after choosing *AB* in the preceding example, we seek the shortest link between *A* or *B*, and *C*, *D*, or *E*; this is *AD*, as before. In the second construction, we begin with the whole graph, and successively remove the longest edge whose deletion does not disconnect the graph. So, in the preceding example:

> We first remove the edges *CE* (length 13), *DE* (length 12), and *BE* (length 11).
>
> We cannot then remove *AE* (length 10), because this would disconnect the graph, so instead we remove *BC* (length 9) and *AC* (length 8).
>
> This leaves the spanning tree that we obtained earlier.

Kruskal concluded his paper with a proof that Construction A always produces the desired minimum spanning tree.

Robert C. Prim received his bachelor's degree in electrical engineering from the University of Texas at Austin, and his doctoral degree in mathematics from Princeton University. During World War II, he worked at the General Electric Company, and later at the US Naval Ordnance Laboratory in Maryland. He spent much of his career at Bell Laboratories, while also working for Sandia National Laboratories in New Mexico. In 1957, while at the Bell Telephone Company, Prim produced his paper on the minimum spanning tree problem, in which he presented a variation on Kruskal's constructions, gave a more substantial example of its use, and explained why his method was more suitable for computer implementation.[69]

R. C. Prim: *Shortest connection networks and some generalizations* (1957)

Prim's paper on the minimum spanning tree problem opens with these words:

> A problem of inherent interest in the planning of large-scale communication, distribution and transportation networks also arises in connection with the current rate structure for Bell System leased-line services. It is the following:
>
> Basic Problem—*Given a set of (point) terminals, connect them by a network of direct terminal-to-terminal links having the smallest possible total length.*

His method for solving this problem involved the joining of "fragments" by links of shortest length, where a *fragment* is a set of terminals (vertices) that are connected by direct links (edges). A terminal or fragment is *isolated* if it has no external connections; for example, the following partial network has three isolated terminals, 2, 4, and 9, and two isolated fragments, 8–3 and 1–6–7–5.

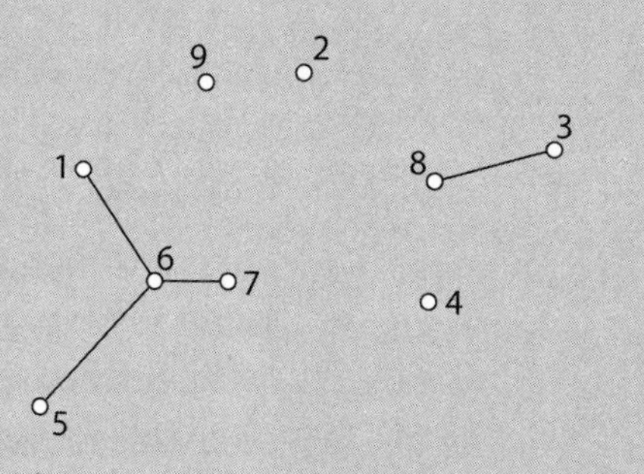

To obtain the required shortest connection network, Prim applied the following two construction principles:

P1: Any isolated terminal can be connected to a nearest neighbor.
P2: Any isolated fragment can be connected to a nearest neighbor by a shortest available link.

For example, in the above network we might proceed as follows:

apply P1 to terminal 4, by adding the link 4–8;
apply P2 to the fragment 4–8–3, by adding the link 8–2;
apply P1 to terminal 9, by adding the link 9–2;
apply P2 to the fragment 1–6–7–5, by adding the link 1–9.

This gives the following shortest connection network.

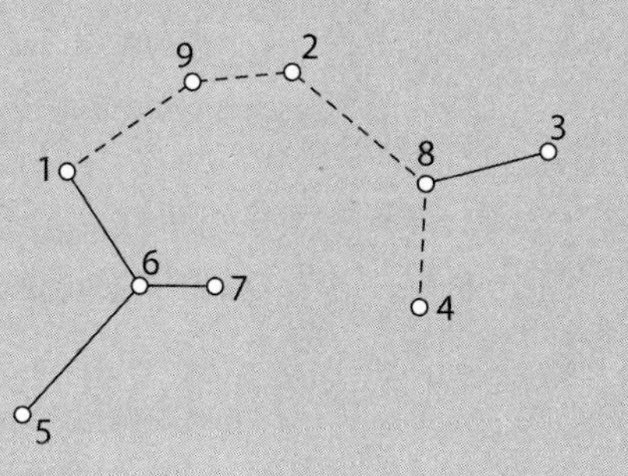

As a second example, Prim considered the network formed by the (then) 48 state capitals and Washington, DC, where the appropriate distances are assumed known.

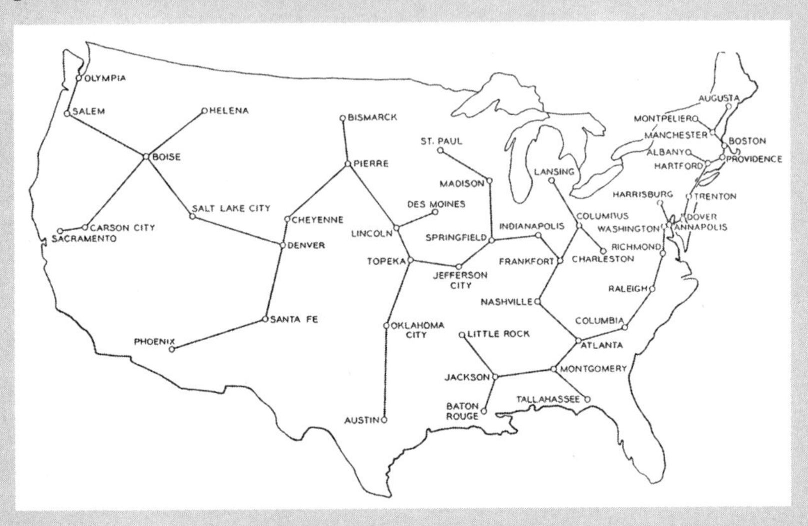

As he observed, the network could be constructed by applying

> P1: Olympia–Salem; P2: Salem–Boise; P2: Boise–Salt Lake City; P1: Helena–Boise; P1: Sacramento–Carson City; P2: Carson City–Boise; P2: Salt Lake City–Denver; P1: Phoenix–Santa Fe; P2: Santa Fe–Denver; and so on,

and he commented that

> With only a few minutes of practice, an example as complex can be solved in less than ten minutes.

Prim then suggested a simpler version, where we use P1 to begin with, and then P2 ever after; this has "advantages for computer mechanization", because we do not have to list all the links and their lengths in advance. Here, starting from Sacramento, we get the following links, where at each step we join a capital already visited to a new one:

> Sacramento–Carson City; Carson City–Boise; Boise–Salt Lake City; Boise–Helena; Boise–Salem; Salem–Olympia; Salt Lake City–Denver; Denver–Cheyenne; Denver–Santa Fe; and so on.

Because the application of either P1 or P2 reduces by 1 the number of isolated terminals and fragments, an N-terminal network becomes connected after $N-1$ applications.

The rest of Prim's paper concerned the justification and computer implementation of his method, and comparisons with Kruskal's approach. He concluded by contrasting the minimum spanning tree problem with two well-known unsolved challenges:

> The simplicity and power of the solution afforded by P1 and P2 for the Basic Problem of the present paper comes as something of a surprise, because there are well-known problems which *seem* quite similar in nature for which no efficient solution procedure is known.
>
> One of these is *Steiner's Problem*: Find a shortest connection network for a given terminal set, with freedom to add additional terminals whenever desired. A number of necessary properties of these networks are known, but do not lead to an effective solution procedure.
>
> Another is the *Traveling Salesman Problem*: Find a closed path of minimum length connecting a prescribed terminal set. Nothing even approaching an effective solution procedure for this problem is now known.

Kruskal and Prim were not the originators of the minimum spanning tree problem, as explained in a historical article on the minimum spanning tree problem by Ron Graham and Pavol Hell.[70] In 1926, the Czech mathematician, Otakar Borůvka, contributed to an engineering magazine a short note on electrical power networks in Western Moravia; in it, he solved an example with forty cities, by linking each city to its nearest neighbor and joining the resulting pairs together by a succession of links of minimum length. Borůvka also wrote a theoretical paper that was cited by both Kruskal and Prim, in which he proved why his method works. An English translation and analysis of both papers was later written by J. Nešetřil et al.[71]

In 1930, Borůvka's compatriot, Vojtěch Jarník, presented an alternative solution, by choosing a city and selecting each successive link so that it joined a city previously visited to a new one. This is the method that was later rediscovered by Prim.[72] Four years later, Jarník collaborated with Milos Kössler on the Steiner problem, mentioned at the end of Prim's paper, where extra locations can be added. In their paper, they proved that such a minimum-length tree always exists, and gave solutions in some special cases.[73]

The other unsolved problem that Prim mentioned at the end of his paper superficially resembles the minimum spanning tree problem:

> *Traveling salesman problem*: Given a number of cities connected by links that join cities in pairs, and given the distances between all pairs of cities, what is the shortest cyclic route that visits every city?

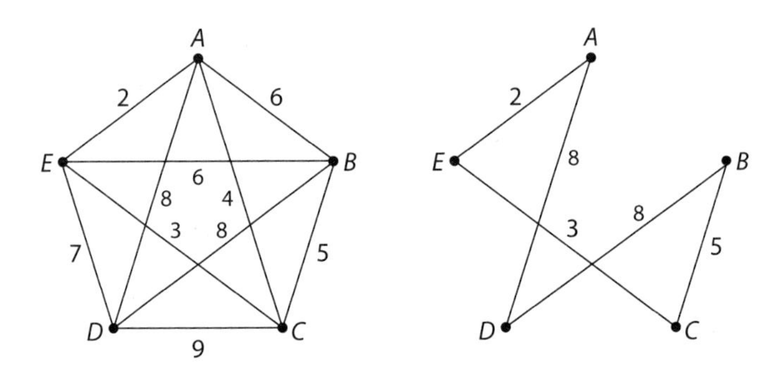

A traveling salesman problem and its solution.

The traveling salesman problem received an early mention in a sales manual of 1831. It was discussed by Karl Menger at a mathematical colloquium in Vienna in 1930, and soon afterward at a seminar talk by Hassler Whitney at Princeton University. After reappearing from time to time in the 1940s, and having been popularized by Merrill Flood of Columbia University, it achieved prominence in 1954 in an influential paper by George Dantzig, Ray Fulkerson, and Selmer Johnson of the RAND Corporation. Here they used linear programming techniques to solve the problem for the (then) 48 US state capitals and Washington, DC.[74] It has since been solved for networks with many thousands of cities. Further information about the traveling salesman problem can be found in several books on the subject.[75]

Search Algorithms and Path Problems

In several practical problems we need to visit every vertex of a graph in a systematic way, and various search algorithms have been developed for doing so. The best known of these are *depth-first search* and *breadth-*

first search, both of which originated in the tracing of mazes—depth-first search, by C. P. Trémaux in the late 19th century, and breadth-first search, by Edward F. Moore in the 1950s.[76] Depth-first search can also be employed when we seek flow-augmenting paths in capacitated networks, whereas breadth-first search can be used for the shortest path problem that we describe below.

In a depth-first search, we penetrate the graph as deeply as possible, before backtracking to other vertices. For example, if we start from vertex A in the following tree, we might proceed to B, and then go directly to D and I. We cannot now go any deeper, so we pick up J and K, before backtracking to A via B, picking up E along the way. We then proceed down the other branch, visiting C, F, and L, before backtracking, picking up G and H on the way back to vertex A. This gives the vertex-ordering

$$A \to B \to D \to I \to J \to K \to E \to C \to F \to L \to G \to H.$$

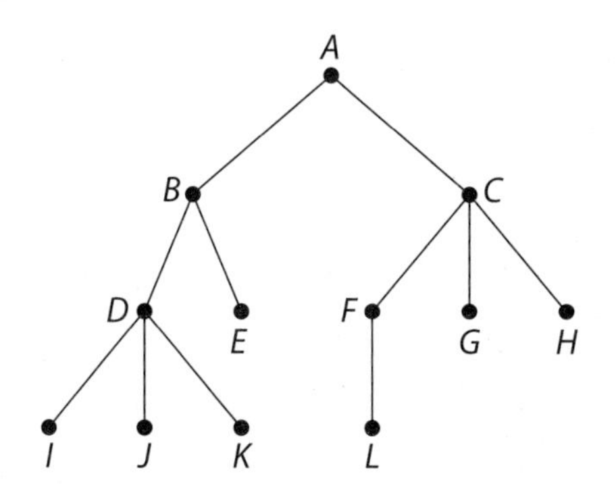

In breadth-first search, we visit all the nearby vertices, before proceeding to more remote ones. For example, starting from vertex A in the above tree we first visit its neighbors B and C, and then the neighbors of B (which are D and E) and of C (which are F, G, and H). Finally, we visit the neighbors of D (which are I, J, and K), and of F (which is L). This places the vertices in alphabetical order:

$$A \to B \to C \to D \to E \to F \to G \to H \to I \to J \to K \to L.$$

The *shortest path problem* is that of determining the shortest route between two given locations in a network or weighted graph. For example, given a road map of the United States, what is the shortest highway route from Los Angeles to Boston?

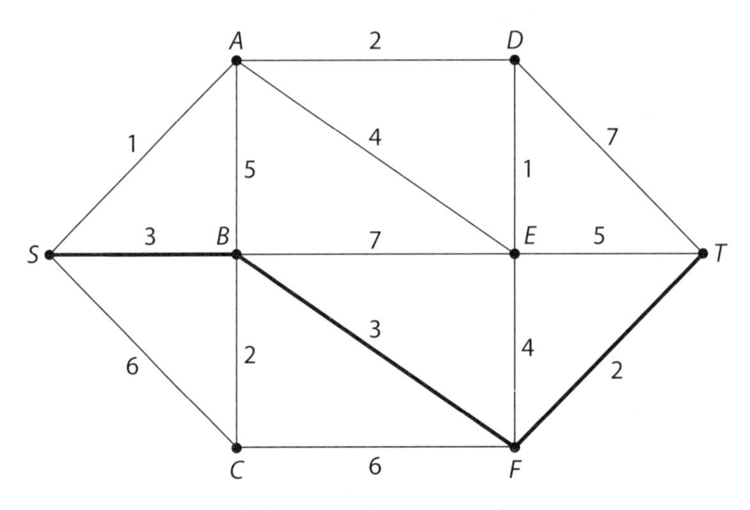

A shortest path in a network.

Several algorithms for finding a shortest path were developed in the 1950s by Richard Bellman, Lester Ford Jr., and others. One imaginative method was presented in 1957 by George J. Minty of the University of Michigan:[77]

> Build a string model of the travel network, where knots represent cities and string lengths represent distances (or costs). Seize the knot 'Los Angeles' in your left hand and the knot 'Boston' in your right hand and pull them apart.

Others used a breadth-first search, beginning with the cities nearest to the start, before moving out to remoter ones; the best-known of these algorithms, published in 1959, was by the Dutch computer scientist Edsger Dijkstra, who applied it to sixty-four locations in the Netherlands.[78] Two years earlier, Dantzig had used linear programming methods to find the shortest route from San Francisco to Boston.[79]

A related problem is the *longest path problem*, where we seek the path of greatest length between a source and a sink. Such problems arise in *critical path analysis*, where we are faced with the scheduling of key tasks in a project (such as the erection of a building), and where certain tasks cannot begin until other tasks have been completed. Here, the situation may be modeled by an "activity network", in which the vertices represent the activities to be carried out (such as laying the drains), and each arc is labeled with the time needed to complete the activity.

The *critical path method* (CPM) for finding such paths was initiated in the late 1950s by Morgan R. Walker of DuPont and James E. Kelley of Remington Rand. Under the acronym of PERT (Program Evaluation

Research Task), it was developed by the US Navy for the analysis and design of large and complex projects, such as the Polaris nuclear submarine. PERT was subsequently renamed "Program Evaluation and Review Technique".

FRANK HARARY

> There are several reasons for the acceleration of interest in graph theory. It has become fashionable to mention that there are applications of graph theory to some areas of physics, chemistry, communication science, computer technology, electrical and civil engineering, architecture, operational research, genetics, psychology, sociology, economics, anthropology, and linguistics. The theory is also intimately related to many branches of mathematics, including group theory, matrix theory, numerical analysis, probability, topology, and combinatorics. The fact is that graph theory serves as a mathematical model for any system involving a binary relation. Partly because of their diagrammatic representation, graphs have an intuitive and aesthetic appeal. Although there are many results in this field of an elementary nature, there is also an abundance of problems with enough combinatorial subtlety to challenge the most sophisticated mathematician.

So begins the preface of *Graph Theory*,[80] one of the most important and frequently cited textbooks in the subject, and one that has defined, developed, and directed the path of modern graph theory. Its author, Frank Harary, was a major and memorable figure on the graph theory stage, both in America and throughout the world. In addition to his writings on mathematical topics, he applied graph theory to many of the subject areas that he listed in the preface. With his wide range of interests and his ever-enthusiastic propagation of the subject, it is not surprising that he came to be known as "the father of modern graph theory".

Frank Harary was born of Syrian and Russian parents and grew up in New York City, being awarded his bachelor's and master's degrees by Brooklyn College. After a year at Princeton University (learning theoretical physics) and a year at New York University (studying applied mathematics), he moved to the University of California, Berkeley, where he was awarded a doctoral degree in 1948 for the thesis *The Structure of Boolean-like Rings*. He then transferred to the University of Michigan in Ann Arbor, where he was appointed as a research assistant in the Institute for Social Research and an instructor in the department of mathematics. He eventually achieved promotion to professor of mathematics in 1964.

Frank Harary (1921–2005) enjoying a combinatorics conference in Balatonfüred, Hungary, in 1969; behind him is Robin Wilson.

Throughout his long career, Frank Harary lectured widely in the United States and in over seventy countries, and spent sabbatical periods in several academic institutions, including Princeton's Institute of Advanced Study and the Universities of London, Oxford, and Cambridge. While at Ann Arbor, he helped to launch the *Journal of Combinatorial Theory* in 1966 and the *Journal of Graph Theory* in 1977. On his retirement from Michigan in 1987, he transferred to New Mexico State University in Las Cruces as Distinguished Professor of Computer Science, a position that he occupied until his death in 2005.

In addition to his eight books, including *Graph Theory*, Harary wrote more than seven hundred papers, including over one hundred that appeared in journals devoted to subjects other than mathematics. His writings covered many areas of graph theory, both pure and applied, of which we outline just three: signed graphs, graph enumeration, and Ramsey graph theory.

Signed Graphs

In connection with his work with the Institute for Social Research at Michigan, Harary introduced the concepts of "signed graphs" and "balance", extending earlier ideas of Dénes König and the psychologist Fritz Heider. A *signed graph* is a graph in which each edge is designated as either *positive* (+) or *negative* (–). A path or cycle in the graph is *positive* if it has an even number of negative edges, and is *negative* otherwise. In the language of social networks, we can think of the vertices as people, and the positive and negative edges as indicating pairs of individuals who are friends or non-friends. For example, a positive 3-cycle corresponds to three mutual friends, or to two friends who have a common non-friend, while a negative 3-cycle corresponds to three mutual non-friends, or to two non-friends who have a common friend.

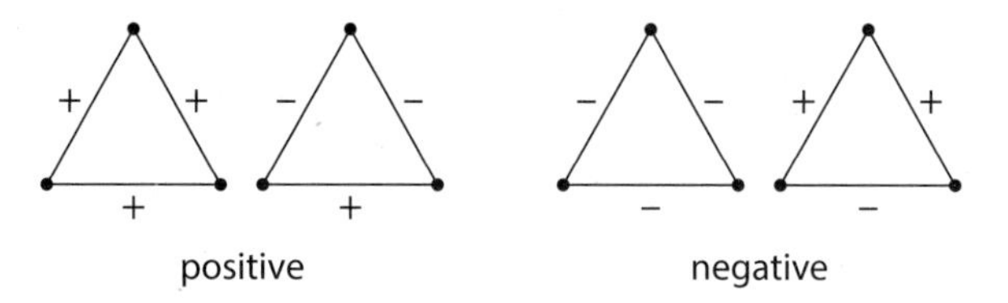

3-cycles in a signed graph.

A signed graph is called *balanced* if every cycle is positive, and in 1954 Harary wrote the first of several papers on balance in signed graphs,[81] proving that

> A signed graph is balanced if and only if its vertices can be partitioned into two disjoint sets (one of which may be empty) in such a way that each positive edge joins vertices in the same set, and each negative edge joins vertices in different sets.

In general, balanced signed graphs correspond to social situations that are stable. With this interpretation, Harary's result asserts that a group of people is stable if it can be split into two parts so that the people within each part are friends, while those in the different parts are not.

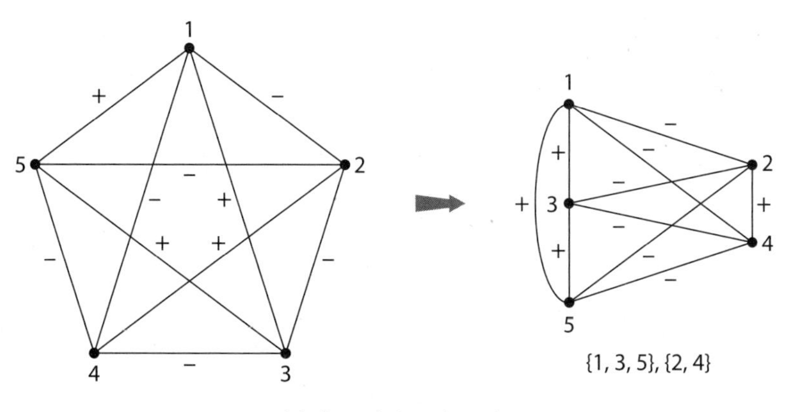

A balanced signed graph.

A few years later, Harary developed these ideas in his book *Structural Models: An Introduction to the Theory of Directed Graphs*, written with his former research student Robert Z. Norman and his colleague Dorwin Cartwright.[82]

Graph Enumeration

As we have seen, graph enumeration is the counting (up to isomorphism) of graphs of a specified kind. Earlier in this chapter, we saw how Arthur Cayley and Richard Otter contributed to the counting of rooted and unrooted trees, and in Chapter 3, we explained how the methods of Redfield and Pólya led to the enumeration of various other types of graphs.

Throughout his life, Harary wrote extensively about enumeration. In 1955, he built on Pólya's work in a significant paper in which he enumerated linear, directed, rooted, and connected graphs, and in the same year he also wrote an elegant note showing how Otter's formula for trees could be deduced directly from Pólya's results.[83] He continued to write papers on the enumeration of graphs and digraphs of various types, and in 1973 he brought all of these contributions together in *Graphical Enumeration*, a book written with his former research student Edgar M. Palmer.[84]

Ramsey Graph Theory

An area of graph theory that developed in the 1950s was *Ramsey graph theory*, named after the British logician and philosopher Frank P. Ramsey, who proved in 1930 that (loosely speaking) subsets of a specified type can always be found within sets that are sufficiently large.[85] In the words of the American mathematician Ron Graham:[86]

> Ramsey theory says that complete disorder is impossible. There is always structure somewhere.

The topic became popular after the following question appeared, at Frank Harary's suggestion, in the annual William Lowell Putnam Mathematical Competition for 1953:

> Six points are in general position in space (no three in a line, no four in a plane). The fifteen line segments joining them in pairs are drawn, and then painted, some segments red, some blue. Prove that some triangle has all its sides the same color.

A similar question had been posed in the Hungarian Mathematical Olympiad of 1947 and frequently appears in books on recreational mathematics in the form:

> If there are six people at a party, prove that there must be at least three mutual friends or three mutual non-friends.

In graphical terms, this problem asks us to show that any red and blue coloring of the edges of the complete graph K_6 must contain either a red K_3 or a blue K_3. To prove this, we select a vertex v and consider the five edges incident to it; at least three of these edges must have the same color—say, va, vb, and vc are all red. Then, if any of the edges ab, ac, or bc is red, we have a red triangle, and if not, then abc is a blue triangle.

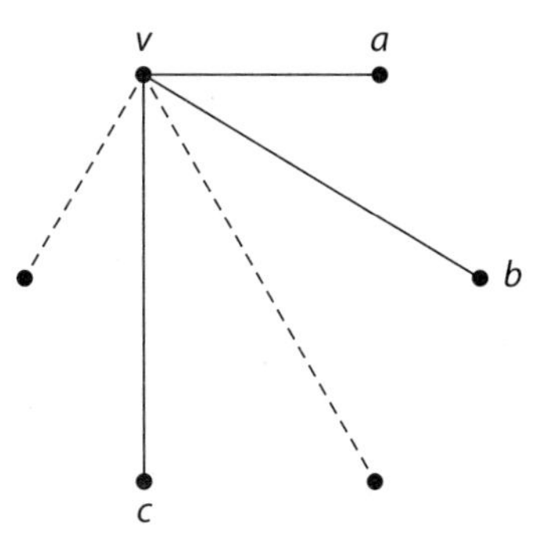

Because the edges of the complete graph K_5 can be colored red and blue without producing a red K_3 or a blue K_3, we have just seen that K_6 is the smallest complete graph that must contain a red K_3 or a blue K_3. We express this by saying that the "Ramsey number $r(3, 3)$ is 6".

More generally, in 1955 Robert E. Greenwood and Andrew M. Gleason defined the *Ramsey number* $N = r(m, n)$ to be the minimum number of vertices that are needed to ensure that any red–blue coloring of the edges of K_N contains either a red K_m or a blue K_n—that is, m mutual friends or n mutual non-friends—and found the following Ramsey numbers:[87]

$$r(4, 3) = 9,\ r(5, 3) = 14,\ r(4, 4) = 18, \quad \text{and} \quad r(2, n) = n, \text{ for any } n;$$

general bounds for $r(m, n)$ can be found in an earlier paper by Paul Erdős.[88] Greenwood and Gleason also extended the problem to a larger number of colors, and proved that $r(3, 3, 3) = 17$—that is, with the three colors red, blue, and green, 17 vertices are needed to ensure a red, blue, or green K_3.

In later years Frank Harary extended the idea much further, in an expanded series of thirteen mainly co-authored papers under the general heading of "generalized Ramsey theory". Here, the complete graph K_3 of the classical Ramsey problems is replaced by other graphs, such as a complete bipartite graph $K_{r,s}$, a path P_n, or a cycle C_n. The *generalized Ramsey number* $r(H_1, H_2)$ is the smallest number N of vertices that are needed to ensure that any red–blue coloring of the edges of K_N contains a red graph H_1 or a blue graph H_2. For example, in one of these papers, Václav Chvátal and Frank Harary found the generalized Ramsey numbers for the graphs H_1 and H_2 with up to four vertices,[89] showing in particular that

$$r(K_{1,3}, C_4)=6, \; r(P_4, P_4)=10, \quad \text{and} \quad r(C_4, K_4)=10.$$

* * * * *

In the twenty years covered by this chapter, graph theory in America and elsewhere was transformed from a rather narrow discipline, primarily obsessed with map coloring and related topics, into a wide-ranging area of mathematics with links to many areas within mathematics and with important applications to most of the sciences and social sciences. This transformation would continue over the next twenty years, as major problems were solved and graph theory increasingly became part of mainstream mathematics.

Chapter 6
The 1960s and 1970s

The 1960s witnessed graph theory becoming ever more established throughout North America and the world, through the publication of new textbooks, the organization of international conferences, and the gradual emergence of the subject in university and college curricula. Applications became widespread, especially in the physical sciences, operational research, and the rapidly developing area of computer science. New names, such as Oystein Ore, Gerhard Ringel and Ted Youngs, and Jack Edmonds, come to feature in our story, and the decade saw the long-awaited proof of the Heawood conjecture on the coloring of maps and graphs on topological surfaces.

These developments continued into the 1970s, with the establishment of regular series of conferences, with the publication of increasingly many books on specific areas of the subject, and with many universities and colleges then presenting graph theory, combinatorics, and operational research among their offerings. Research was active in a wide variety of areas, both theoretical and practical, as the subject continued to gain credence. Most noteworthy of all was the publicity arising from Kenneth Appel and Wolfgang Haken's proof of the four color theorem—the subject's most celebrated challenge—which had been a central preoccupation for over a hundred years.

OYSTEIN ORE

Øystein Ore (pronounced "oo-reh", and usually spelled in the United States as "Oystein") was born in 1899 in Norway's capital city of Kristiania (later renamed Oslo). He graduated from the University of Kristiania in 1922, and two years later he received his doctoral degree for a thesis on the theory of algebraic fields. He then traveled around Europe, and visited the University of Göttingen, the Mittag-Leffler Institute in

Oystein Ore (1899–1968).

Sweden, and the Sorbonne in Paris, before he returned as a research assistant at what was by then the University of Oslo. His mathematical interests at this time lay mainly in algebraic number fields, rings, and Galois theory, with graph theory, lattice theory, and the history of mathematics emerging as later concerns.

In 1927, Ore was recruited by Yale University and moved there as an assistant professor. He became a full professor in 1929 and two years later was promoted to Sterling Professor, Yale's highest academic position. He was a plenary speaker on algebra at the International Congress of Mathematicians in Oslo in 1936 and remained at Yale until his retirement in 1968, just before his untimely death.

During World War II, Ore became actively involved with the "American relief for Norway" and "Free Norway" movements. In 1947, King Haakon VII decorated him with the Royal Norwegian Order of Saint Olav, an honor that eighteen years earlier had been awarded to Oswald Veblen.

Oystein Ore wrote nine books while at Yale, on topics ranging from algebra and number theory to the history of mathematics. In 1936, Dénes König had written his groundbreaking text on graph theory (see Interlude B), and no other books on the subject then appeared until 1958, with the publication of Claude Berge's *Théorie des Graphes et ses Applications*, followed by an English translation in 1962.[1] In the summer of 1941, Ore had presented the American Mathematical Society's Colloquium Lectures on "Mathematical relations and structures" in Chicago, but these were not published at the time. They eventually appeared in 1962 in the AMS Colloquium Lecture series, as his *Theory of Graphs*.[2] These books by Berge and Ore were the first texts on the subject to be published in the English language.

Ore wrote two further books on graph theory. In 1963, his book *Graphs and Their Uses*[3] introduced the subject to high school students. Four years later, his classic *The Four-Color Problem*[4] was the first major book in English devoted exclusively to the topic.

Ore wrote few papers on graph theory, but one of them seems particularly noteworthy. One would expect that graphs with many edges are more likely to be Hamiltonian than those with fewer edges, and several general results on Hamiltonian graphs are of this type. For example, Gabriel Dirac proved in 1952 that if a graph has n vertices, and if the degree of each vertex is at least $n/2$, then the graph is Hamiltonian.[5] Ore's note, remarkably just one page in length, presents a strengthening of Dirac's theorem.[6]

Oystein Ore: *Note on Hamilton circuits* (1960)

In his note, Ore proved that if G is a graph with n (≥ 3) vertices, and if

$$\deg(v) + \deg(w) \geq n,$$

whenever the vertices v and w are not adjacent, then G is Hamiltonian. Dirac's theorem is clearly a special case of this result.

We recall that if G is Hamiltonian, then we can arrange its vertices in cyclic order,

$$a_0, a_1, a_2, \ldots, a_{n-1}, a_0,$$

where a_k is adjacent to a_{k+1} for each k, and a_{n-1} is adjacent to a_0. It follows that, if G is not Hamiltonian, then any cyclic arrangement of its vertices must have "gaps", where a gap occurs between a_k and a_{k+1} when they are not adjacent. To obtain a contradiction, Ore assumed that G is not Hamiltonian, and also, without loss of generality, that the preceding cyclic order has a minimum number of gaps, one of which he took to be between a_0 and a_1.

He next observed that if a_0 is adjacent to a_i, for some i, then a_1 cannot also be adjacent to a_{i+1}, because otherwise we could reorder the vertices as

$$a_0, a_i, a_{i-1}, \ldots, a_2, a_1, a_{i+1}, a_{i+2}, \ldots, a_{n-1}, a_0,$$

with one fewer gap, as shown below.

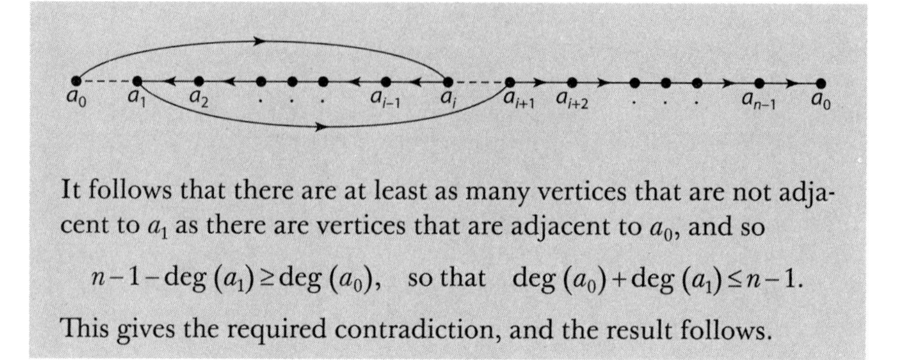

It follows that there are at least as many vertices that are not adjacent to a_1 as there are vertices that are adjacent to a_0, and so

$$n-1-\deg(a_1)\geq\deg(a_0), \quad \text{so that} \quad \deg(a_0)+\deg(a_1)\leq n-1.$$

This gives the required contradiction, and the result follows.

One of Ore's doctoral students at Yale was Marshall Hall Jr., who became a well-known combinatorialist. Another was Joel G. Stemple, who received his doctorate in 1966 and subsequently worked with Ore on the four color problem. Together they were able to extend C. E. Winn's result from almost thirty years earlier (see Chapter 3), announcing their achievement in the *Notices of the American Mathematical Society*:[7]

> OYSTEIN ORE, Yale University, New Haven, Connecticut, and JOEL STEMPLE, Queens College, City University of New York, Flushing, New York 11367. On the four color problem. It is shown that a planar map not colorable in four colors must have at least n = 40 countries. This improves on the result n = 36 due to C. E. Winn (1938). The rather elaborate calculations are based upon the Euler contributions of the faces in an irreducible graph and upon three new reducible configurations. (Received October 13, 1967.)

Their proof appeared in 1970,[8] but it involved so many special cases that the full details could not be published and were lodged in the library of Yale's mathematics department. A mistaken assumption in their proof was subsequently discovered and quickly dealt with by Frank Bernhart. Jean Mayer, a professor of French literature at the University of Montpellier and keen amateur mathematician, then managed to push the number of countries in an irreducible map up to 48,[9] and later to 96—but there was still a long way to go.

THE HEAWOOD CONJECTURE

> In 1890 P. J. Heawood . . . published a formula which he called the Map Colour Theorem. But he forgot to prove it. Therefore the world of mathematicians called it the Heawood Conjecture. In 1968 the formula was proven

Gerhard Ringel (1929–2008) and Ted Youngs (1910–70) in 1968.

> and therefore again called the Map Color Theorem. (This book is written in California, thus in American English.)

So commented Gerhard Ringel in the foreword to his classic book *Map Color Theorem*,[10] which tells the story of how he and J.W.T. Youngs proved the Heawood conjecture, with assistance from several others. We next examine some of the ideas which led to the *tour de force* that is now known as the *Ringel–Youngs theorem*.

Gerhard Ringel

Gerhard Ringel was born in Austria and raised in Czechoslovakia, where he graduated from Charles University in Prague. During World War II, he was drafted into the Wehrmacht, later spending four years as a prisoner of war in a Soviet jail. On his release, he attended the University of Bonn, where he received his doctoral degree in 1951 and taught for a further nine years. In 1960, he was appointed professor of mathematics at the Free University of Berlin, and ten years later he accepted a permanent position at the University of California in Santa Cruz, succeeding his friend Ted Youngs, who had retired.

We recall from Interlude A the various equivalent forms of the Heawood conjecture. For orientable surfaces:

> *Heawood conjecture* (*map coloring*): For each $g \geq 1$, the chromatic number of the orientable surface S_g is $\chi(S_g) = \lfloor ½(7 + \sqrt{1+48g}) \rfloor$.

Heawood conjecture (*neighboring regions*): The simplest orientable surface on which n (≥ 3) neighboring regions can be drawn is S_g, where $g = \lceil \frac{1}{12}(n-3)(n-4) \rceil$.

Heawood conjecture (*complete graphs*): The orientable genus of $K_n (n \geq 3)$ is

$$g(K_n) = \lceil \tfrac{1}{12}(n-3)(n-4) \rceil.$$

We note that the formula for the chromatic number of S_g also happens to give the correct answer, $\chi(S_0) = 4$, for the number of colors needed for maps drawn on the sphere ($g = 0$). But we cannot deduce the four color theorem from a proof of the Heawood conjecture, which applies only when $g \geq 1$.

For non-orientable surfaces, noting the exceptional case of the Klein bottle in every case (see Chapter 3):

Heawood conjecture (*map coloring*): For each $q \geq 1$, the chromatic number of the non-orientable surface N_q is $\chi(N_q) = \lfloor \frac{1}{2}(7 + \sqrt{1 + 24q}) \rfloor$, except that $\chi(N_2) = 6$.

Heawood conjecture (*neighboring regions*): The simplest non-orientable surface on which n (≥ 3) neighboring regions can be drawn is N_q, where $q = \lceil \frac{1}{6}(n-3)(n-4) \rceil$, except that $q = 3$ when $n = 7$.

Heawood conjecture (*complete graphs*): The non-orientable genus of K_n ($n \geq 3$) is

$$\hat{g}(K_n) = \lceil \tfrac{1}{6}(n-3)(n-4) \rceil,$$

except that $\hat{g}(K_7) = 3$.

From the early 1930s, rumors were circulating that the Heawood conjecture had been proved. For example, in the first edition of Richard Courant and Herbert Robbins's well-known book, *What Is Mathematics?*, published in 1941, it was claimed that:[11]

> for surfaces more complicated than the plane or sphere the corresponding theorems have actually been proved.

But these reports were premature, and it was not until the 1950s that much progress would be made.

The breakthrough was achieved by Gerhard Ringel. In his doctoral thesis for the University of Bonn, he investigated coloring theorems for non-orientable surfaces, and followed this with a remarkable sequence of papers involving both the orientable and non-orientable cases.[12] In particular, he solved the complete graphs problem for orientable surfaces when $n = 8$, and for all values of n of the form $12s + 5$, such as 17, 29, 41, etc.

In 1954, Ringel published a paper that completely settled the neighboring regions and complete graphs versions of the Heawood conjecture for all *non-orientable* surfaces.[13] It may seem surprising that these surfaces should have been easier to deal with than their orientable analogs (the many-holed toruses), but Ringel found an argument that avoided splitting the problem into many separate cases, as would later be necessary for the orientable version.

In 1959, Ringel produced a significant book on the coloring of graphs, *Färbungsprobleme auf Flächen und Graphen* (Coloring Problems for Surfaces and Graphs).[14] This book and his many papers were major contributions to topological graph theory.

Before we outline the history of the solution to the Heawood conjecture for orientable surfaces, we make some preliminary remarks. Two versions of the conjecture involve the number 12, and the eventual proof turned out to split into twelve separate cases, depending on the remainder when n (the number of neighboring regions or vertices) is divided by 12—that is, on the value of n (modulo 12). If n is of the form $12s + k$—that is, $n \equiv k \pmod{12}$—this is called *Case k*. In 1891, Heffter had already solved the neighboring regions problem for several values of n in Case 7 (see Interlude A), and we have just seen that Ringel later solved Case 5 in its entirety.

Of particular interest are Cases 0, 3, 4, and 7, because for each of these, $\frac{1}{12}(n-3)(n-4)$ is an integer that does not need to be "rounded up". In these cases, the corresponding embedding of the complete graph K_n on the surface turns out to be a *triangulation*, where every region is a triangle. For the other eight cases, the solution involved two parts: the *regular part*, where a "near-triangulation" was found, and the *additional adjacency problem*, where the embedding had to be adjusted to cope with the particular situation. For some of these cases, the additional adjacency problem proved to be the most difficult part, involving the addition of an extra handle or some other adjustment.

For a brief overview of the methods employed, we now look at the simplest case, Case 7. Here $n = 12s + 7$, for some s, and we seek an embedding of the complete graph K_n on the appropriate orientable surface; as we just mentioned, this embedding is a triangulation of the surface.

Case 7

In 1891, one of Heffter's most fruitful contributions to the proof of the Heawood conjecture had been the introduction of *rotation schemes* that specify, for each region of a map, the colors of its neighboring regions, listed in counter-clockwise order. In its dual version, a rotation scheme specifies, for each vertex of the complete graph K_n, the colors of its neighboring vertices, listed in cyclic order. To see what is involved, consider the following drawing of K_7 on a torus.

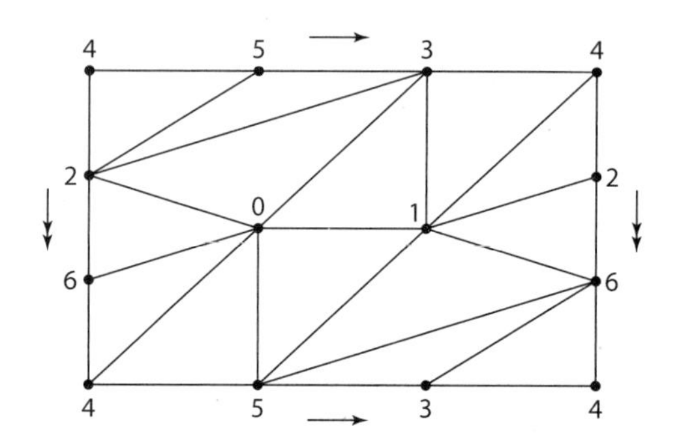

Here, the vertex labeled 0 is adjacent to the vertices labeled 1, 3, 2, 6, 4, and 5 in counter-clockwise order, and we record this by writing

(0): 1 3 2 6 4 5

as the first row of our rotation scheme. Repeating this for the vertex that is labeled 1, we have

(1): 2 4 3 0 5 6.

Continuing in this way, we obtain the following rotation scheme, which describes the arrangement of the vertices. Notice how each successive row is obtained from the previous one by adding 1 (modulo 7) to each number, so that the entire coloring is determined, once we have found the first row:

(0): 1 3 2 6 4 5
(1): 2 4 3 0 5 6
(2): 3 5 4 1 6 0
(3): 4 6 5 2 0 1
(4): 5 0 6 3 1 2
(5): 6 1 0 4 2 3
(6): 0 2 1 5 3 4.

Conversely, given a rotation scheme, we can assign to any vertex the label 0, and then label the other vertices accordingly. This gives the required triangular embedding.

But how do we find the first row of the rotation scheme, from which the other rows can then be generated? In 1963, William Gustin of Indiana University and the National Bureau of Standards presented an ingenious way of doing so. He considered an associated network, which he called a *current graph*, of arcs labeled with numbers that satisfy Kirchhoff's current laws.[15]

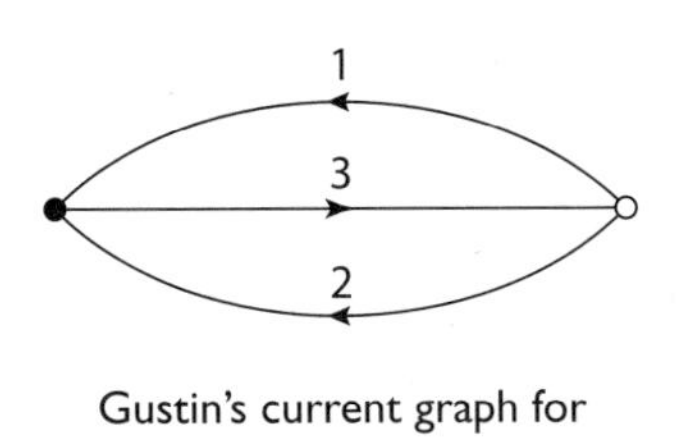

Gustin's current graph for $n=7$.

To see what is involved, consider the current graph for $n=7$, and follow this rule:

> *Rule*: Follow the arcs, turning *left* at each solid vertex (•) and *right* at each hollow vertex (○), and list the arcs as you go, with a minus sign whenever you need to traverse an arc in the wrong direction.

So, following this rule, and starting with arc 1 and listing the arcs as we go, we have:

follow arc 1 to vertex •	1
turn left, and follow arc 3 to vertex ○	3
turn right, and follow arc 2 to vertex •	2
turn left, and follow arc 1 backwards to vertex ○	–1
turn right, and follow arc 3 backwards to vertex •	–3
turn left, and follow arc 2 backwards to vertex ○	–2

Every arc has now been traced in both directions, and we have the sequence

1 3 2 –1 –3 –2,

which (modulo 7) is

1 3 2 6 4 5.

This is the first row of the rotation scheme, from which the other rows can then be deduced.

The next example in Case 7 is $n=19$, and here we use the following current graph:

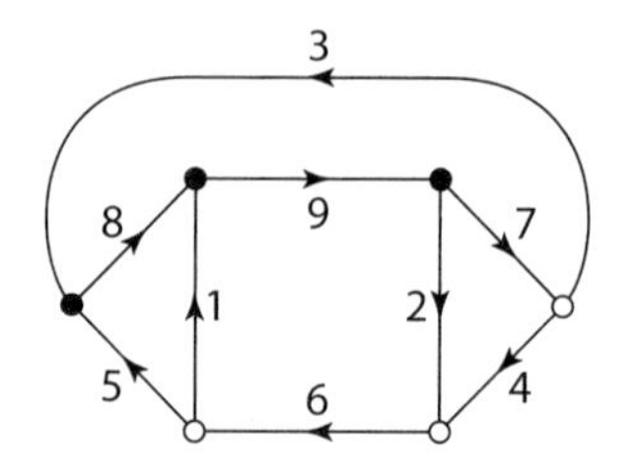

Gustin's current graph for $n=19$.

To generate the rotation scheme, we carry out the preceding rule as before, again starting with arc 1:

follow arc 1 to vertex ●	1
turn left, and follow arc 8 backwards to vertex ●	–8
turn left, and follow arc 5 backwards to vertex ○	–5
and so on.	

When every arc has been traced in both directions, we have the sequence

1 –8 –5 –6 –4 3 8 9 7 4 –2 –9 –1 5 –3 –7 2 6,

which (modulo 19) is

1 11 14 13 15 3 8 9 7 4 17 10 18 5 16 12 2 6.

This yields the following rotation scheme, in which each successive row is obtained by adding 1 (modulo 19) to the preceding row:

(0):	1	11	14	13	15	3	8	9	7	4	17	10	18	5	16	12	2	6
(1):	2	12	15	14	16	4	9	10	8	5	18	11	0	6	17	13	3	7
(2):	3	13	16	15	17	5	10	11	9	6	0	12	1	7	18	14	4	8
								.....										
(18):	0	10	13	12	14	2	7	8	6	3	16	9	17	4	15	11	1	5,

from which the labeling of the vertices can be determined.

For the general case, when $n=12s+7$, Gustin used the following current graph to find the rotation scheme, and thereby the triangular embedding of K_n.

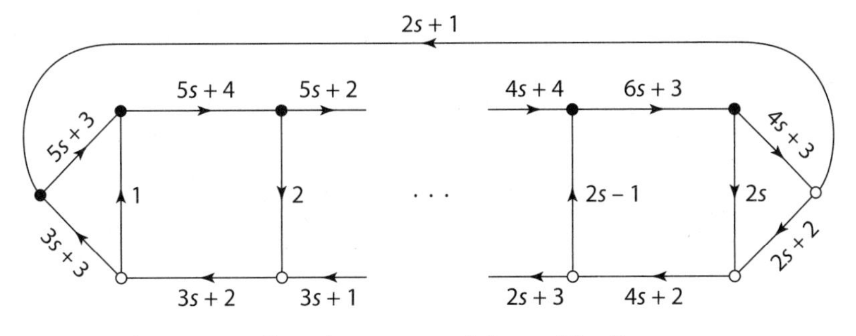

Gustin's current graph for $n = 12s + 7$.

The Ringel–Youngs Theorem

John William Theodore (Ted) Youngs was an American, born in India but educated in the United States, who gained his doctoral degree in 1934 from the Ohio State University for a thesis on topology. He then taught at Ohio State and Purdue Universities, served in the US Air Force during World War II, and was a consultant to the Institute of Defense Analysis and to industry. After the war, he taught at Indiana University for eighteen years, moving in 1964 to the new Santa Cruz campus of the University of California, where he remained until his death in 1970. His numerous papers on topology contained many significant results, with his best early work on the abstract concept of a surface, but today he is mainly remembered for his contributions to graph theory during the last ten years of his life.

Between 1963 and 1968, Ringel and Youngs collaborated on several papers related to the Heawood conjecture. They, and others, had dealt with nine of the twelve cases, but the remaining three—Cases 2, 8, and 11—caused particular difficulties, and Youngs invited Ringel to spend the academic year 1967–68 on sabbatical in California trying to resolve them. Their success in doing so resulted in the paper, "Solution of the Heawood map-coloring problem",[16] published in 1968, which brought together all their findings.

Gerhard Ringel and J.W.T. Youngs, *Solution of the Heawood map-coloring problem* (1968)

This three-part paper began with a description of the four color problem, Kempe's attempted solution, Heawood's discovery of Kempe's error, and a description of the various forms of Heawood's conjecture and of the connections between them.

The second part of the paper summarized how the solution was reached:

1891: Heffter solved the complete graphs problem for $n \leq 12$, and for certain values of n in Case 7.

1952: Ringel proved the equivalence of Heawood's conjecture on the chromatic number of a surface and the neighboring regions problem, and found the orientable genus of K_{13}.

1954: Ringel solved Case 5, the first case to be settled completely.

1961: Ringel solved Cases 7, 10, and 3.

1962: Gustin became interested in Cases 0, 3, 4, and 7, introduced current graphs, and announced solutions to Cases 3, 4, and 7, although his example for Case 4 was in error.

1963: C. M. Terry, L. R. Welch, and Youngs found solutions to Case 4 and Case 0; Gustin also solved Case 4.

1963–64: Youngs developed a theory of "vortex graphs", and with Gustin solved Case 1.

1965: Case 9 was solved, mainly by Gustin.

1966: Youngs solved Case 6.

1967–68: Ringel and Youngs joined forces in California and solved Cases 2, 8, and 11, except for a few small values of n: 18, 20, 23, 30, 35, 47, and 59. These were solved by various people, including Jean Mayer of the University of Montpellier.

The last part of this paper illustrated a solution of Case 8, when $n=32$. Here the regular part of the problem used a current graph to construct a map with 33 regions on a surface, and the additional adjacency part added an extra handle to the surface in order to present a map in which all 32 regions are mutually adjacent.

With the help of others, Ringel and Youngs had indeed proved the Heawood conjecture for orientable surfaces. Two years later, Ted Youngs died, and Ringel recalled:[17]

> I can only say that working together proved to be extremely profitable and enjoyable for both of us. Ted's enthusiasm, his knowledge, and his dedication to this problem as well as to mathematics, in general, were admirable.

The solutions for all twelve cases were published in the *Journal of Combinatorial Theory*, and in Ringel's subsequent book, *Map Color Theorem*, which appeared in 1974.

We conclude this section with a true story, which shows that solving map coloring problems can yield unexpected benefits:[18]

> Shortly after the Heawood conjecture had been proved, Ringel's wife was driving along the California expressway, but was stopped by a traffic cop for a traffic violation. On learning that her name was "Ringel", the traffic cop inquired: "Is your husband the one who solved the Heawood conjecture?" Surprised, she gave an affirmative answer, and was duly let off with only a warning. It transpired that the traffic cop's son had been in Ted Youngs's calculus class when the proof of the conjecture had been announced.

RON GRAHAM

No book on discrete mathematics in America would be complete without a mention of one of its most colorful characters. Once described as "one of the charismatic figures in contemporary mathematics, as well as the leading problem-solver of his generation",[19] Ronald Lewis Graham wrote around 400 papers on a wide variety of topics that ranged from graph theory and computational geometry to number theory and the theory of approximation. But as well as being remembered as a popular writer and lecturer who enjoyed juggling mathematical symbols, he had also been a circus performer who juggled balls and clubs and became president of the International Jugglers' Association.[20]

Ron Graham had an unconventional childhood. His father worked in oil fields and shipyards, frequently changing jobs and working variously in California, Georgia, and Florida. As a result, Ron rarely remained at any school for longer than a year or so, and he never graduated from high school. His parents eventually divorced, and while living with his mother in Florida, he was awarded a Ford Foundation scholarship at the age of just 15 to study at the University of Chicago. Although he had already developed an interest in astronomy and mathematics, he spent much of his time in Chicago in gymnastic activities, such as acrobatics and trampolining. When his scholarship ran out, he spent a year at the University of California, Berkeley, majoring in electrical engineering while also taking a one-year course on number theory with D. H. Lehmer and continuing with his acrobatics.

In 1955, in order to avoid being drafted, he enlisted for four years in the US Air Force and was sent for three of these years to Fairbanks,

Ronald L. Graham (1935–2020).

Alaska. In addition to his military duties as a telephone operator, which he fulfilled at night, he enrolled at the University of Alaska and graduated in 1959 with a bachelor's degree in physics. Returning to Berkeley, he received a master's degree in 1961 and a doctoral degree a year later for a thesis on number theory, supervised by Derrick Lehmer. To fund his studies at Berkeley, he created a trampolining troupe (the Bouncing Baers) and a juggling act (the Fumbling Franklins), and performed in circuses and elsewhere.

In 1962, Ron Graham began to work for Bell Laboratories in Murray Hill, New Jersey (later renamed AT&T Laboratories), where he became the director of information services, managing the research activities of a large department; he stayed there for 37 years. In 1982 his responsibilities as head of the Mathematical Studies Center were described as follows:[21]

> For the past 20 years he has confronted the formidable challenges that arise from the need to route hundreds of millions of telephone calls through the intricate communications web of cables, microwaves, and satellites that embraces the earth. The mathematical techniques and theorems he has developed in the process can be applied not only to the routing of information with a computer, but also to the efficient scheduling of an astronaut's day, or even to the allocating of an entire nation's resources . . . By finding better approaches to the traveling salesman and other related problems, Graham and his colleagues continue to improve long-line telephone efficiency.

He had the rare ability to translate real-world problems into mathematics and, in the words of a mathematical colleague from Stanford University:[22]

> Ron, as much as anybody, is responsible for bringing high powered math to bear on computer science.

While at Bell Laboratories, Ron Graham developed a world-class center for research in many areas of discrete mathematics. From his first two years, when he completed eight papers, he gradually built up to producing a dozen or more research publications per year. In particular, among those papers on graph theory that fall within our time period was an important pair with his Bell Labs colleague Henry O. Pollak that related to telephone switching theory.[23] As Pollak later recalled:[24]

> It started when John Pierce, my boss, walked into my office and said: "I've invented a new system for data transmission, and everything is clear except how the messages are going to figure out where to go. See if you can figure something out." That's how that particular problem started. It turned out to be a very exciting graph-theory development. We had to consider the properties of the distance matrix of the graph. People had only studied the adjacency matrices of graphs before. The distance matrices of graphs had fantastically interesting properties that were a solution to this addressing problem that John was interested in. It has led to some very nice mathematics.

Graham and Pollak's main theorem had wide-ranging consequences. One of these is an evaluation of the determinant of the distance matrix $D=(d_{i,j})$ of a tree, where $d_{i,j}$ is the length of the shortest path between vertices i and j:[25]

> For a tree T with n vertices,
>
> $$\det(D) = -(n-1)(-2)^{n-2},$$
>
> and is independent of the structure of T.

An ingenious proof of this result is based on a determinant rule discovered by the English mathematician C. L. Dodgson (better known as Lewis Carroll, the author of *Alice's Adventures in Wonderland*).

Another consequence of the Graham–Pollak theorem involves splitting a complete graph into edge-disjoint complete bipartite graphs:[26]

> The edges of the complete graph K_n cannot be partitioned into fewer than $n-1$ complete bipartite graphs.

This result is best possible, because we can number the vertices 1, 2,..., n and successively join each vertex to all higher-numbered vertices, producing a partition into exactly $n-1$ subgraphs of the form $K_{1,k}$. For example, we can split the complete graph K_6 into the five subgraphs with edges

(12, 13, 14, 15, 16), (23, 24, 25, 26), (34, 35, 36), (45, 46), and (56).

Another partition splits K_6 into the complete bipartite graphs $K_{3,2}$, $K_{2,2}$, $K_{1,3}$, $K_{1,1}$, and $K_{1,1}$.

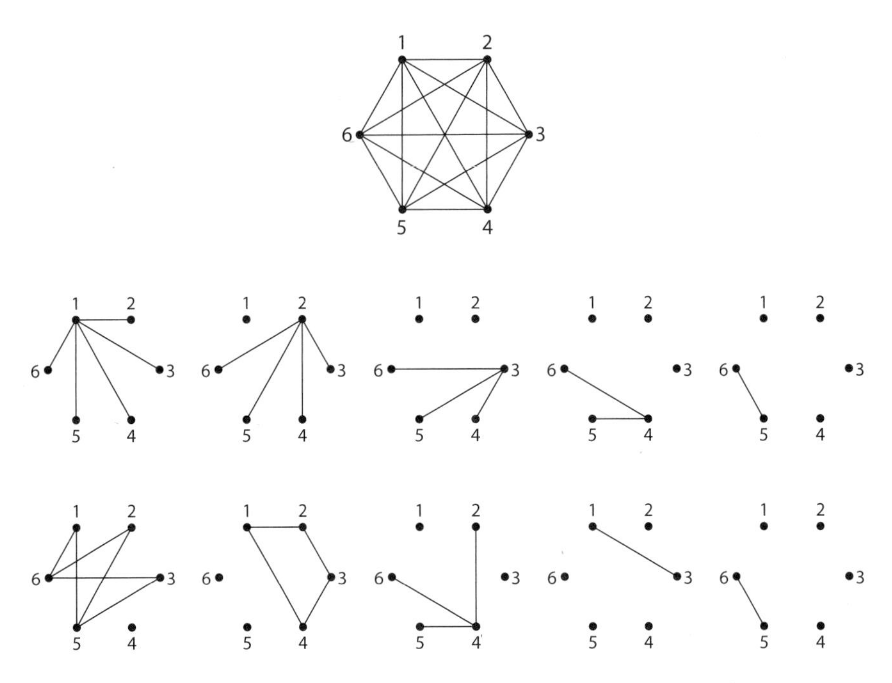

Two partitions of K_6 into edge-disjoint complete bipartite graphs.

Another topic in which Graham had a great interest, and on which he wrote extensively, was Ramsey theory (see Chapter 5).[27] He co-authored an early result in 1975 with his Bell Labs colleague Fan Chung (who later became his wife, and with whom he wrote around 100 papers). It concerned the generalized Ramsey number $r(C_4, C_4, \ldots, C_4)$; where there are k copies of the cycle graph C_4, k colors are available, and we seek the smallest number N of vertices that are required to ensure that any coloring of the edges of K_N contains a copy of C_4 in one of the k colors. In 1975, they proved that:[28]

$$r(C_4, C_4, \ldots, C_4) \le k^2+k+1, \text{ for all } k,$$
$$\text{and } r(C_4, C_4, \ldots, C_4) > k^2-k+1 \text{ if } k \text{ is a prime-power.}$$

Graham's activities were also closely connected with those of his friend, the itinerant and prolific Hungarian mathematician Paul Erdős; they first met at a conference in Boulder, Colorado, in 1963 and wrote many joint papers. In later years, he used to organize Erdős's life, arranging his academic visits, booking his travel tickets and hotels, buying his clothes, and paying his taxes. He also defined the *Erdős number* of an author to be the length of the shortest chain of collaborators from Erdős to the author; for example, if A has written a joint paper with Erdős, then A's Erdős number is 1; if not, but if A has written a joint paper with B who has written a joint paper with Erdős, then A's Erdős number is 2, and so on for larger numbers.

Ron Graham won many honors and awards and was frequently seen with his juggling equipment at various meetings where he was giving prestigious lectures. He is one of the few mathematicians who became president of both the Mathematical Association of America and the American Mathematical Society. From the latter organization he was awarded the Leroy P. Steele Award for Lifetime Achievement with the following citation, which summarizes his wide-ranging accomplishments:[29]

> Ron Graham has been one of the principal architects of the rapid development worldwide of discrete mathematics in recent years. He has made many important research contributions to this subject, including the development, with Fan Chung, of the theory of quasirandom combinatorial and graphical families, Ramsey theory, the theory of packing and covering, etc., as well as the theory of numbers, and seminal contributions to approximation algorithms and computational geometry (the "Graham scan"). Furthermore, his talks and his writings have done much to shape the positive public image of mathematical research in the USA, as well as to inspire young people to enter the subject. He was chief scientist at Bell Labs for many years and built it into a world-class center for research in discrete mathematics and theoretical computer science.

On receiving the award, Graham characteristically replied:

> I can't remember a time when I didn't love doing mathematics, and that desire has not dimmed over the years (yet!). But I also get great pleasure sharing mathematical discoveries and insights with others, even though this

can present a special challenge for mathematicians talking to nonmathematicians. However, I really believe that this type of communication will become increasingly important in the future.

COMPLEXITY

In 1798, in his *Essay on the Principle of Population*, Thomas Malthus contrasted the linear rise in food supply with the competing exponential growth of the population. He predicted that, however well people may survive in the short term, the exponential increase would eventually dominate, with severe consequent food shortages—a conclusion that was borne out in practice.

Malthus's perception of the difference between linear (or, more generally, polynomial) growth and exponential growth became of the greatest importance in the 1950s and 1960s. With the rapid advance of computer science, it became increasingly necessary to distinguish between algorithms that were efficient, and those that were likely to take much longer to carry out.

But what do we mean by "efficient"? Consider the minimum spanning tree problem and the traveling salesman problem that we described in Chapter 5:

> *Minimum connector problem*: A railway network is to be built connecting several towns. If the distances between all pairs of towns are known, how can we design the network so that the total amount of railway track is as small as possible?
>
> *Traveling salesman problem*: Given a number of cities connected by links that join cities in pairs, and given the distances between all pairs of cities, what is the shortest cyclic route that visits every city?

For the former problem, we presented algorithms which rapidly produce the shortest spanning tree that connects all the cities; these "greedy algorithms" are easy to describe and quick to carry out. But for the second problem, if there are N cities, then there are $(N-1)!$ cyclic routes that pass through all the cities, and this number grows very fast as N increases. It seems unlikely that there is an efficient algorithm for this problem—and indeed, no such algorithm has ever been found.

All algorithms have a *running time*, such as the maximum time that a computer requires to complete all the necessary calculations or the number of such calculations. Every problem also has an *input size*, which may be the number of cities in a network or the number of vertices in an associated graph. The running time generally increases with the input size, but exactly how quickly this occurs is often the crucial question.

Particularly important are *polynomial-time algorithms*, in which the running time is proportional to some power of the input size n—often, to n^2 or n^3. For any number k, we say that the running time of a graph algorithm is $O(n^k)$ if it is at most a constant multiple of n^k. Several of the graph problems in Chapter 5 have polynomial-time algorithms, where the input size n is the number of vertices. Examples are the following:

the minimum connector problem, where the running time is $O(n^2)$,
the shortest path problem, where the running time is $O(n^2)$,
the maximum matching problem, where the running time is $O(n^3)$,
the maximum flow problem, where the running time is $O(n^3)$.

But there are also algorithms that usually take longer than polynomial time—such as the *exponential-time algorithms* with running time proportional to an nth power, such as 2^n or 10^n. To see the difference between polynomial-time and exponential-time algorithms, let us compare two algorithms with running times n^3 and 2^n on a computer that performs one million operations per second. If the input size n is 10, then

$$10^3 = 1000 \quad \text{and} \quad 2^{10} = 1024,$$

and so both algorithms have running times of about 0.001 second. But if the input size n is 50, then

$$50^3 = 125{,}000 \quad \text{and} \quad 2^{50} = 1{,}125{,}899{,}906{,}842{,}624,$$

and the former algorithm has a running time of 0.125 second, whereas the latter algorithm takes 35.7 years to complete.

So, in general, exponential-time algorithms take significantly longer to carry out than polynomial-time ones. Algorithms that run in polynomial time are generally thought to be "efficient", whereas those that run in exponential time are considered "inefficient", certainly as the input size increases. In practice, however, an algorithm whose running time is n^{100} may not be faster than one with running time $2^{n/100}$, unless n is very large (say, $n = 200{,}000$). Moreover, a running time that is proportional to n^3 might exceed one that is proportional to 2^n, for small values of n, if the constant of proportionality of the former greatly exceeds that of the latter. For example, if $100n^3$ is compared with $2^n/100$, then the polynomial-time algorithm becomes faster only when n is at least 27.

Interest in the efficiency of algorithms began to increase in 1953, when John von Neumann distinguished between polynomial-time and exponential-time algorithms in a paper linking a particular zero-sum two-person game to the optimal assignment problem.[30] The dichotomy was further discussed in the mid-1960s by Alan Cobham of IBM's Research Center at Yorktown Heights, New York, and by several others.[31]

Jack Edmonds (b. 1934).

Particularly important contributions were made by the American-born Jack Edmonds, who spent much of his working life in Canada, in the Department of Combinatorics and Optimization at the University of Waterloo. After receiving his master's degree from the University of Maryland for a dissertation on the embedding of graphs on surfaces, he worked at the National Bureau of Standards and the RAND Corporation in California. In 1965, his award-winning article "Paths, trees, and flowers"[32] built on W. T. Tutte's criterion for the existence of a perfect matching (1-factor) in a graph (see Chapter 5). It presented a remarkable polynomial-time algorithm for finding a maximum matching in a general graph. Edmonds followed this with other seminal work, which extended his results to weighted graphs and linked them to linear programming and related topics.[33] He also worked on matroids and proved a major result called the "matroid intersection theorem", which links matroids to minimax theorems, linear programming, and duality.[34]

The so-called "Cobham–Edmonds thesis", named after Alan Cobham and Jack Edmonds, characterized "easy, fast, and practical" computational problems as those that can be solved by polynomial-time algorithms, or "good algorithms", as Edmonds called them. However, in his 1965 paper, Edmonds also observed that:[35]

> One can find many classes of problems, besides maximum matching and its generalizations, which have algorithms of exponential order but seemingly

> none better. An example known to organic chemists is that of deciding whether two given graphs are isomorphic. For practical purposes the difference between algebraic and exponential order is often more crucial than the difference between finite and non-finite.

Before we move on from polynomial algorithms, we mention just one more type of problem—that of determining whether a given graph with n vertices is planar. Kuratowski's criterion involving the graphs K_5 and $K_{3,3}$ (see Interlude B) does not provide a suitable method. More successful was an approach by L. Auslander and S. V. Parter,[36] who successively removed cycles from the graph and examined the resulting subgraphs for planarity; this led to an algorithm with running time $O(n^3)$. But the *coup de grâce* was made in 1974, with the publication by John Hopcroft and Robert Tarjan of a linear algorithm—one with running time $O(n)$—in their seminal paper, "Efficiency planarity testing".[37]

The P versus NP Problem

The collection of all tractable problems—those that can be solved by a polynomial-time algorithm—is denoted by P. But, as we have seen, there are many problems for which no polynomial algorithm has ever been found; these include the traveling salesman problem. However, if someone were to propose a possible cyclic route for a salesman to take, then we can certainly check in polynomial time whether that route is cyclic, and whether its length is less than the minimum length already known.

At this point, we introduce NP, the set of "non-deterministic polynomial-time problems". These are problems for which a solution, when proposed, can be *checked* in polynomial time; it follows that the traveling salesman problem is in NP. Indeed, the introduction of NP was partly due to Jack Edmonds, who had earlier conjectured that the traveling salesman problem has no polynomial solution.

Clearly, P is contained in NP, because if a problem can be solved in polynomial time then a solution can certainly be checked in polynomial time—checking solutions is usually easier than finding them in the first place—and this leads naturally to a fundamental question that may have originated with Edmonds around 1967:

$$\text{Is } P = NP?$$

Equality seems very unlikely—certainly, few people believe that the answer to this question is "yes", because this would mean that all problems for which a solution is easy to check would also be easy to solve. But this has never been proved.

Stephen Cook and NP-completeness

Important insight into the "P versus NP" problem was provided by Stephen A. Cook, another American-born mathematician and computer scientist who moved to Canada. After graduating from the University of Michigan in 1961 and receiving his doctoral degree at Harvard University in 1966 for the thesis *On the Minimum Computation Time of Functions*, Cook taught at the University of California at Berkeley from 1966 to 1970. When his position there was not renewed, he transferred to the University of Toronto, where he remained.

In 1971, Cook wrote a seminal paper, "The complexity of theorem-proving procedures",[38] in which he considered a particular NP problem in mathematical logic called the *satisfiability problem*. Here, the problem is to determine whether there is an assignment of truth values (true/false) to the variables that makes a given logical expression true: for example, the expression

$$(a \vee e \vee i) \wedge (\bar{a} \vee e \vee \bar{i}) \wedge (a \vee \bar{e} \vee o) \wedge (\bar{a} \vee \bar{i} \vee \bar{o})$$

(where the variables $\bar{a}$, $\bar{e}$, $\bar{i}$, and $\bar{o}$ are the negations of a, e, i, and o) is true when a and e are true and i and o are false.

Cook proved the startling result that any other problem in NP can be transformed into the satisfiability problem in polynomial time. It follows

Stephen Cook (b. 1939).

that if the satisfiability problem is in P, then so also is everything else in NP and we can deduce that P = NP, but if the satisfiability problem is not in P, then P ≠ NP. So the whole P versus NP question depends on whether there is a polynomial algorithm for solving just one particular problem.

Based on these findings, Cook defined a problem to be NP-*complete* if its solution in polynomial time implies that *every* NP problem can be solved in polynomial time. These NP-complete problems include the satisfiability problem and hundreds of others, such as the following:

the traveling salesman problem;

the Hamiltonian cycle problem: does a given graph have a Hamiltonian cycle?

the 3-colorability problem: can the vertices of a given graph be colored in three colors?

the isomorphism problem: are two given graphs isomorphic?

A discussion of such problems appears in two survey articles by Richard Karp;[39] it was in the first of these that the notations P and NP were introduced. More information can be found in Michael Garey and David Johnson's classic text, *Computers and Intractability: A Guide to the Theory of NP-Completeness*,[40] published in 1979, which includes a comprehensive list of more than three hundred NP-complete problems. If a polynomial algorithm could be discovered for any one of them, then polynomial algorithms would exist for them all, and P would equal NP. If, on the other hand, it could be proved that any one of these problems has no polynomial algorithm, then none of the others could have a polynomial algorithm either, and P would be different from NP.

Knowing whether P = NP is not just a theoretical matter. Many NP-complete problems are of great practical importance, and an enormous amount of money is at stake, including a millennium prize of one million dollars offered by the Clay Mathematics Institute for deciding the issue. But since the 1970s, little progress has been made in settling this general problem.

It is also worthy of note that several winners of the prestigious A. M. Turing Award of the Association for Computing Machinery—Stephen Cook (1982), Richard Karp (1985), and John Hopcroft and Robert Tarjan (1986)—have all featured in our story. Indeed, graph theory has played a central role in the general development of the theory of algorithms and their complexity. Not only does algorithmic theory use examples from graph theory, but such examples and applications have been fundamental for creating some of its basic concepts.

THE FOUR COLOR THEOREM

For over twenty-five years, following the contributions of Philip Franklin, Clarence Reynolds, and C. E. Winn in the 1920s and 1930s (see Chapter 3), little progress had been made in America on solving the four color problem. Interest revived in the mid-1960s, however, partly though the publication of Oystein Ore's book *The Four-Color Problem*, and within just ten years the problem was finally conquered.[41] Meanwhile, progress had already been made in Germany that would have major consequences for the eventual solution.

Heinrich Heesch and Wolfgang Haken

Heinrich Heesch (pronounced "haish") was born in Kiel, Schleswig-Holstein, in 1906. After graduating in mathematics and music from the University of Munich, and receiving his doctoral degree from the University of Zürich for a dissertation on geometry, he moved to Göttingen University where he assisted Hermann Weyl with work on the geometry of crystals. Heesch soon achieved some recognition for constructing a tiling of the plane that answered a problem presented by David Hilbert; Hilbert had included it among the twenty-three problems that he posed at the 1900 International Congress of Mathematicians in Paris, thereby setting the agenda for mathematical research activity in the 20th century.

In 1935, because of what was happening with the National Socialists and the Nazi work camps, Heesch resigned his university post and returned to Kiel. Here, he earned his living as a schoolteacher until he was able to gain employment at the Leibniz University Hannover (later the Technische Universität of Hannover) and the University of Kiel.

Heinrich Heesch became interested in the four color problem around 1935. He recalled that an unavoidable set of configurations is a collection of configurations, at least one of which must appear in any map, and that a reducible configuration is a configuration for which any coloring of the surrounding ring of regions can be extended to the regions inside (so that it cannot feature in a minimal counter-example to the four color theorem), and gradually came to believe that the best approach to solving the problem would be to search for an *unavoidable set of reducible configurations.* If such a set were to exist, then every map must include at least one of them, and yet none could appear in a counter-example to the theorem, and so the four color theorem would be true. By this time, unavoidable sets of configurations had been presented by Kempe, Wernicke, Franklin, and Lebesgue, and reducible configurations had been produced by Birkhoff, Franklin, Errera, Reynolds, and Winn. But Heesch feared that the required unavoidable set would have to be very large, possibly amounting to many thousands of reducible configurations.

In 1948, Heesch lectured on the four color problem at the University of Kiel to an audience that included the young Wolfgang Haken. Haken recalls Heesch mentioning that ten thousand special cases may need investigation. Five hundred of these configurations had already been checked at a rate of one per day, and Heesch seemed optimistic about working through the remaining nine-and-a-half thousand.

Haken also attended lectures on topology, where he learned of three long-standing unsolved problems: the *knot problem*, of determining whether any given tangle of string contains a knot; the *Poincaré conjecture*, concerning the classification of spheres in four-dimensional space; and the *four color problem*. Haken explored the first of these for his doctoral degree, awarded by the University of Kiel in 1953, and soon after managed to solve the problem in full—a magnificent achievement that was rewarded by an invitation to lecture at the 1954 International Congress of Mathematicians in Amsterdam.

Haken's solution of the knot problem so impressed Bill Boone, a logician at the University of Illinois at Urbana–Champaign, that Haken was invited there as a visiting professor. He then spent a couple of years at the Institute for Advanced Study in Princeton, before returning to a permanent position in Illinois. There he worked on the Poincaré conjecture, which he reduced to the consideration of two hundred special cases. He managed to deal with no fewer than 198 of these, but after struggling with the remaining two for over ten years, he eventually gave up and turned his attention to the four color problem. The Poincaré conjecture was eventually proved by Grigori Perelman in 2004.

Wolfgang Haken (b. 1928).

Unavoidable Sets and Reducible Configurations

In order to set the scene for Haken's work on the four color theorem, we need to say more about unavoidable sets and reducible configurations, and this takes us back to the work of Heinrich Heesch.

In the 1960s, Heesch developed a method for testing unavoidable sets of configurations, which Haken later called the *method of discharging*. We illustrate the idea by explaining why Wernicke's set of configurations—a digon, a triangle, a square, two adjacent pentagons, and a pentagon adjoined to a hexagon (see Chapter 3)—is an unavoidable set.

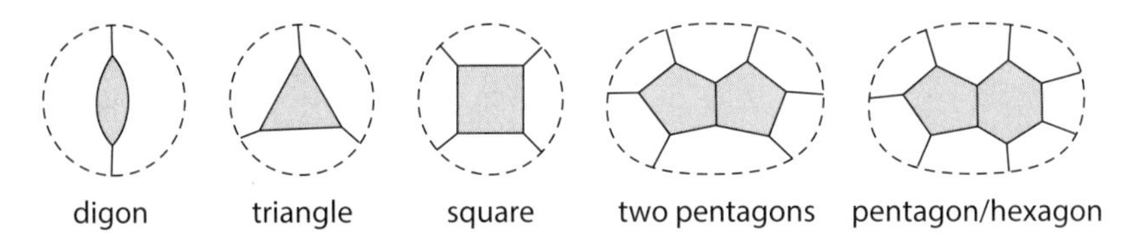

Wernicke's unavoidable set.

Let us assume that Wernicke's configurations do not form an unavoidable set, so that there is a cubic map that includes none of them; in such a map, every pentagon can adjoin only regions with at least seven sides. By the Counting theorem given in Chapter 3,

$$4C_2 + 3C_3 + 2C_4 + C_5 - C_7 - 2C_8 - 3C_9 - 4C_{10} - \cdots = 12,$$

where, for each k, C_k is the number of k-sided regions in the map. It follows that if there are no digons, triangles, or squares, then

$$C_5 - C_7 - 2C_8 - 3C_9 - 4C_{10} - \cdots = 12.$$

We next assign to each region with k sides an "electric charge" of $6 - k$, so that each pentagon receives a charge of 1, hexagons have zero charge, heptagons have charge −1, and so on. It follows from the preceding equation that the total charge of the regions of the map is 12, a positive number. We then move the charges around the map in such a way that no charge is created or destroyed—this is called *discharging the map*.

One way of moving the charges is to transfer one-fifth of a unit of charge from each pentagon to each of its five negatively charged neighbors with seven or more sides. The result of this operation is that each pentagon now has zero charge, each hexagon keeps its zero charge, and (as can easily be checked) no region with seven or more sides receives enough contributions of $\frac{1}{5}$ to acquire positive charge. So all regions of the map now have zero or negative charge, but the total charge is still 12. From this contradiction, we deduce that Wernicke's set of configurations is indeed an unavoidable set.

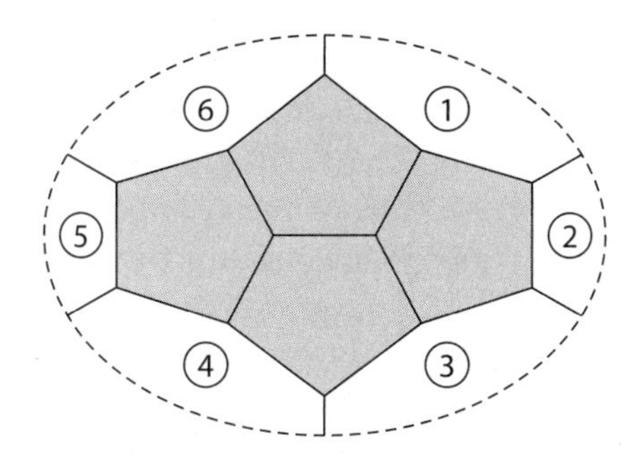

The Birkhoff diamond.

Let us now turn our attention to reducible configurations. In Chapter 2, we claimed that the "Birkhoff diamond", consisting of four pentagons surrounded by a ring of six regions, is a reducible configuration, in the sense that any coloring of the regions in the ring can be extended, either directly or after some Kempe-interchanges of color, to the pentagons inside. For, if the colors are red (r), blue (b), yellow (y), and green (g), then it can be shown that there are thirty-one essentially different colorings of the regions in the ring, such as *rgrgrb* and *rgrbry*. Sixteen of these (such as *rgrgrb*) are "good colorings", in that they can be extended directly to the pentagons inside the ring, whereas the others (such as *rgrbry*) cannot. But by suitable Kempe-interchanges of color, these latter colorings can all become good colorings, and so the Birkhoff diamond is reducible.

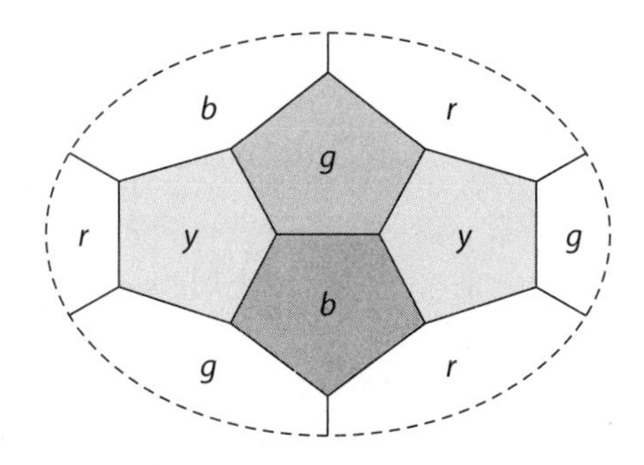

Extending the good coloring *rgrgrb* to the pentagons inside the ring.

Heesch defined a configuration to be *D-reducible* if every coloring of the regions in the surrounding ring extends, either directly or after a succession of Kempe-interchanges, to the interior regions—so a digon, a triangle, a square, and the Birkhoff diamond are all *D*-reducible. He also defined a configuration to be *C-reducible* if it can be proved reducible after it has been modified in some way, and he explored various methods for doing so. We shall return to these concepts shortly.

By this time, almost everyone working on the four color problem was using the dual formulation, first introduced by Kempe, of coloring the vertices of the corresponding planar graph, with adjacent vertices colored differently. In particular, Heesch introduced a notation that became widely used, in which he represented each vertex by a "blob" to make it more easily distinguishable; for example, vertices of degree 5, 6, 7, and 8 (corresponding to pentagons, hexagons, heptagons, and octagons) were represented respectively by ●, •, ○, and □. *In this section, we shall continue with the original version for maps.*

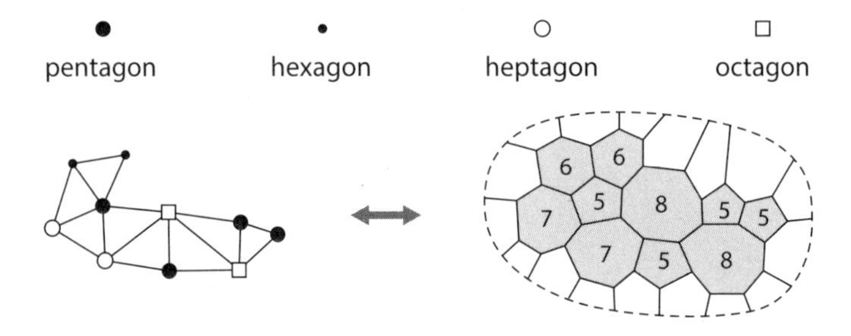

A Heesch drawing and a corresponding configuration of regions.

In 1969, Heesch published a paperback book[42] in which he presented his method of discharging and many of his other contributions to the solution of the four color problem.

Enter the Computer

In 1967, Wolfgang Haken contacted Heinrich Heesch—who by this time had invented the methods of discharging and had discovered thousands of reducible configurations—and invited him to lecture at the University of Illinois. During Heesch's visit, Haken asked him whether computers might help with the analysis of large numbers of configurations. Heesch had already been thinking along similar lines and, with the help of Karl Dürre (a mathematics graduate from Hannover), had developed a method for testing *D*-reducibility that was sufficiently algorithmic to be carried out on the University of Hannover's CDC 1604A computer. By the end of 1965, Dürre had verified that the Birkhoff diamond is

D-reducible and had confirmed the *D*-reducibility of many configurations of increasing complexity.

The complexity of a configuration is usually measured by its *ring-size*, the number of regions surrounding the configuration. As we have seen, the Birkhoff diamond has ring-size 6, with thirty-one essentially different colorings to be examined. Unfortunately, this number grows rapidly as the ring-size increases, as we can see from the following:

ring-size:	6	7	8	9	10	11	12	13	14
colorings:	31	91	274	820	2461	7381	22,144	64,430	199,291

Heesch and Dürre discovered that the time that their computer took to analyze a configuration grew rapidly as the ring-size increased, with a typical configuration with ring-size 12 taking six hours, and those of ring-size 14 being then beyond reach. They also estimated that it might take many thousands of hours to examine all ten thousand cases that Heesch had predicted, and this was unrealistic for any computer of the time. Meanwhile, Edward F. Moore of the University of Wisconsin had discovered large and complicated maps that contained no known reducible configurations, and had deduced that any unavoidable set of reducible configurations must include at least one configuration with ring-size 12 or more.[43]

It was becoming clear that the Hannover computer could no longer carry out the work required of it, and Haken bid to the University of Illinois for time on a new parallel supercomputer whose construction was nearing completion. This was not yet ready for use, but the university's computer department arranged for Heesch and Dürre to use the Cray Control Data 660 computer, the most powerful machine of its day, which was located at the Atomic Energy Commission's Brookhaven National Laboratory in Long Island, New York. Helpfully, Yoshio Shimamoto, the computer center's director, was an enthusiast for the four color problem and strongly supported Heesch's approach to solving it. He invited Heesch and Dürre to visit Brookhaven for extended periods of time to continue their reducibility tests on the Cray computer. As a result, they were able to confirm the *D*-reducibility of over a thousand configurations with ring-size up to 13, and began to test some with ring-size 14.

Meanwhile, Shimamoto was continuing with his own research into the problem and, unlikely though it may seem, showed that if he could find a single *D*-reducible configuration with certain particular properties, then the four color theorem would follow—so the entire proof would depend on just one configuration! On September 30, 1971, he discovered what he was seeking—a configuration with ring-size 14 with the required properties. This configuration became known as the *Shimamoto horseshoe*, and it remained only to confirm that it was indeed *D*-reducible.

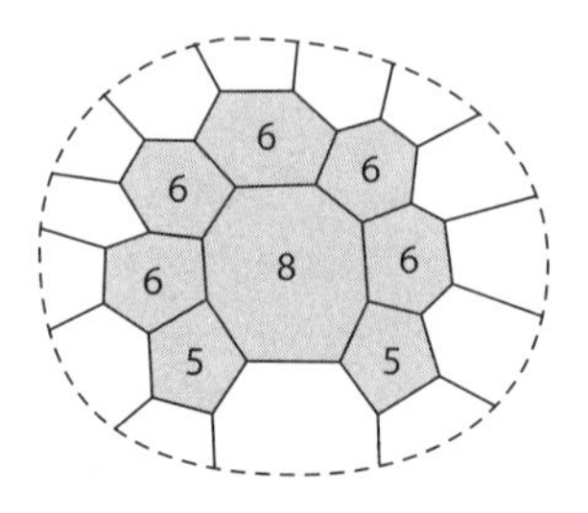

Shimamoto's horseshoe configuration.

By chance, Heesch and Haken were visiting Brookhaven at that time. Heesch believed the horseshoe to be one of his *D*-reducible configurations, and excitement ran riot as rumors about Shimamoto's discovery began to spread around the world. But things did not work out as he had hoped. Three lengthy computer runs to test the *D*-reducibility of the horseshoe were attempted, and in the last of these the Cray computer was allowed to run for a whole weekend. After grinding on for twenty-six long hours, it eventually showed, to everyone's great disappointment, that the horseshoe was *not D*-reducible.

Haken then reworked Shimamoto's theoretical arguments and found them to be completely correct—and this was confirmed in a lengthy and influential paper from an unexpected source. Hassler Whitney and Bill Tutte, the most distinguished graph-theorists of their day, had also "found no essential flaw in Shimamoto's reasoning", and concluded that the computer implementation must therefore be incorrect. They explained Shimamoto's approach as follows:[44]

> Shimamoto, on the assumption that the Four Colour Conjecture was false, showed that there must be a non-colourable map *M* containing a configuration *H* [the horseshoe] that had already passed the computer test for *D*-reducibility. He then arrived at a contradiction by showing that the *D*-reducibility of *H* implied the 4-colourability of *M* . . . The burden of proof was not now on a few pages of close reasoning, but on a computer!

They also recalled how they had independently greeted the method of proof with "some misgivings and then with real scepticism", adding:

> It seemed to both of us that if the proof was valid it implied the existence of a much simpler proof (to be obtained by confining one's attention to one

small part of *M*), and that this simpler proof would be so simple that its existence was incredible. The present paper is essentially the result of our attempts to give a proper mathematical form to our objection.

Whitney and Tutte then presented a detailed discussion of Kempe chains, elementary reductions, and *D*-reducibility, and explained at some length why the Shimamoto horseshoe (and indeed *any* configuration that was produced by his method) could not be *D*-reducible.

It seemed as though the four color problem had again reached a dead end. But Heesch's approach was beginning to pay dividends, and in the hands of Appel and Haken would yield the eagerly desired goal within the next five years.

A New Approach

In 1970, Heesch sent Haken the results of a new discharging experiment in which the positive charge on each pentagon was then to be distributed *equally* to all neighboring regions with negative charge. He estimated that this process would have the effect of reducing the problem to the consideration of 8900 "bad configurations", extending up to ring-size 18, which he proposed to work through one at a time.

By this time, Heesch had developed the uncanny knack of being able to look at a configuration and to predict, with at least 80 percent accuracy, whether it was reducible. He invited Haken to collaborate with him, and in 1971 sent him three "obstacles to reducibility", whose presence seemed to prevent configurations from being reducible. These were:

a 4-*legger region*, which adjoins four consecutive regions of the surrounding ring (marked with stars);

a 3-*legger articulation region*, which adjoins three regions of the surrounding ring that are not all adjacent;

a *hanging* 5–5 *pair*, a pair of adjacent pentagons that adjoin a single region inside the surrounding ring.

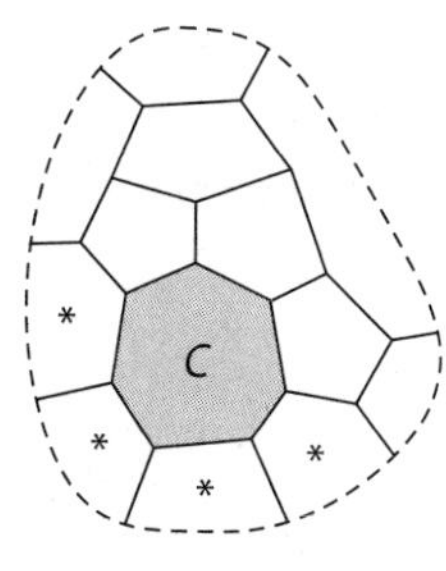

4-legger region

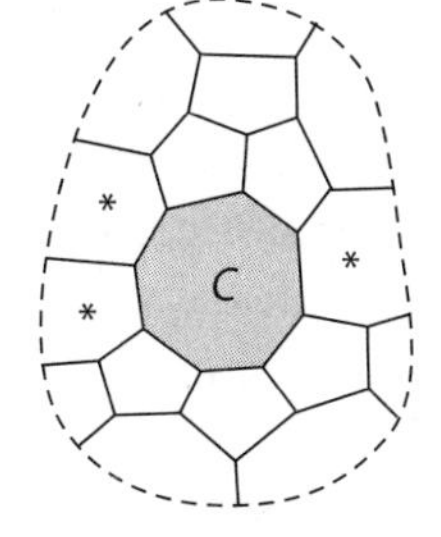

3-legger articulation region

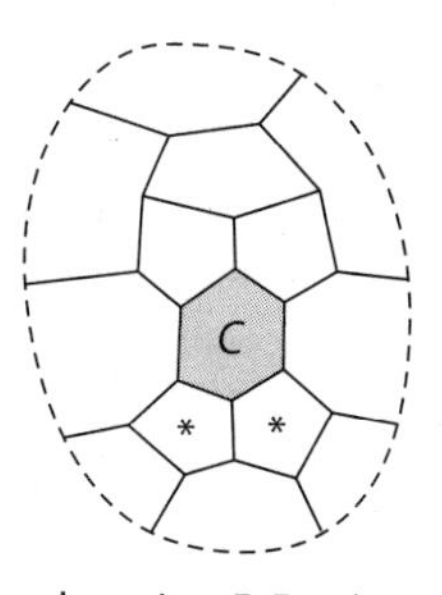

hanging 5-5 pair

But Haken was becoming increasingly pessimistic about being able to deal with so many thousands of configurations, and began to change his approach to the problem. Unlike everyone else's method, which seemed to involve generating reducible configurations by the hundreds and attempting to package them up into an unavoidable set, Haken's primary motivation (which he later developed with Appel) involved aiming directly for an unavoidable set which contained configurations that seemed *likely to be reducible*—in particular, they should not include any of the three obstacles to reducibility—in order to avoid wasting time checking configurations that would eventually be of no interest. Any configurations that subsequently proved not to be reducible could then be dealt with individually as necessary. As he later commented:[45]

> If you want to improve something, you should not improve that part which is already in good shape. The weakest point is always the one you should improve. This is a very simple answer to why we got it and not the others.

So, from this point on, Haken headed in a different direction from everyone else: he concentrated on the unavoidable set and left all details of the reducibility until much later.

Even so, having little knowledge of computing, Haken considered giving up the problem for a few years until more powerful computers were available to deal with the massive calculations that would clearly be necessary. He had been told that his ideas could not be programmed, and during a lecture that he gave in Illinois on the horseshoe episode,[46] he remarked that

> The computer experts have told me that it is not possible to go on like that. But right now I'm quitting. I consider this to be the point to which and not beyond one can go without a computer.

Among those attending Haken's lecture was Kenneth Appel, who had graduated from Queens College, New York, before receiving his doctoral degree from the University of Michigan for a thesis on the application of mathematical logic to problems in algebra. As an experienced computer programmer, he had learned to program at the University of Michigan. He then gained further experience with Douglas Aircraft and the Institute for Defense Analyses at Princeton, before settling at the University of Illinois at Urbana–Champaign. His experience was to prove invaluable for solving the four color problem.

After the lecture, Appel told Haken that he considered the computer experts to be talking nonsense, and offered to work on the problem of implementing the discharging procedures for the creation of unavoidable sets:[47]

> I don't know of anything involving computers that can't be done; some things just take longer than others. Why don't we take a shot at it?

Coincidentally, Appel was a member of the thesis examination panel for a research student of Haken's who had just submitted his doctoral thesis on a special case of the four color problem. Hence, the collaboration could prove beneficial to all concerned.

Haken was delighted to accept Appel's offer to take care of the computing side of things, and they agreed to concentrate their search on unavoidable sets, without taking time to check the configurations for reducibility. In particular, they focused on *geographically good configurations*—those that contained neither of Heesch's first two obstacles to reducibility: such configurations could easily be identified by computer or indeed by hand. They would then check their configurations for reducibility once the entire set had been constructed.

Getting Down to Business

When they started work in late 1972, Appel and Haken's first computer runs already provided much useful information. But the computer output was enormous, with many configurations appearing multiple times: it would be necessary to control these duplications if the eventual list were to be manageable. Fortunately, the computer program had run in just a few hours, and so they would be able to experiment as much as they needed to.

Kenneth Appel (1932–2013).

The necessary changes to the program proved to be straightforward to implement, and the next runs were a definite improvement. As major problems were overcome, lesser ones would emerge, and from then on, every two weeks or so, Appel and Haken modified the discharging algorithm or the computer program as necessary. This two-way dialogue with the computer continued, and as each successive difficulty was sorted out, new ones arose. Within six months of experimenting and improving their procedures, they came to realize that their goal of producing a finite unavoidable set of geographically good configurations in a reasonable amount of time was indeed feasible.

At this stage, they decided to prove *theoretically* that their approach would provide such an unavoidable set: for this they would have to include every possible case, even if it were unlikely to occur in practice. This proved to be more complicated than they had expected, but the eventual outcome, in the fall of 1974, was a lengthy proof that *an unavoidable set of geographically good configurations exists*, together with an achievable method for constructing such a set.[48]

In early 1975, they introduced Heesch's third obstacle to reducibility. This inevitably involved further changes in procedure, but was carried out successfully with only a doubling in the size of the unavoidable set. They also programmed the computer to search for sets of configurations with relatively small ring-size. At this stage, the computer started to think for itself, as Appel and Haken later recalled:[49]

> It would work out complex strategies based on all the tricks it had been "taught" and often these approaches were far more clever than those we would have tried. Thus it began to teach us things about how to proceed that we never expected. In a sense it had surpassed its creators in some aspects of the "intellectual" as well as the mechanical parts of the task.

As soon as it seemed probable that they could find an obstacle-free unavoidable set of configurations that were likely to be reducible, it would be time for them to start their massive detailed check on reducibility. Inevitably, there would be some awkward configurations in the list, but they hoped that these would be relatively few in number. With configurations that might extend to ring-size 16, or cause trouble in other ways, they expected to have to find some clever shortcuts.

In the middle of 1974, realizing that they needed help with the reducibility programs, Appel visited the computer science department to ask whether any graduate student would be interested in joining them. John Koch had just learned that the thesis problem on which he was working had been solved by someone else, and was seeking another topic.

He was quickly set to work on the *C*-reducibility of configurations, where the configuration in question could be modified so that the reducibility arguments would go through, but where it was not always clear how this might be done. Appel and Haken were particularly interested in modifications that could easily be implemented, and Koch discovered that most configurations with ring-size 11 were of this kind. Arguing that little would be gained by including the others, which would have been difficult to program, he focused on simple modifications, devising an elegant method for testing the *C*-reducibility of all configurations with ring-size 11. Appel was then able to extend this method to configurations with ring-sizes 12, 13, and 14.

Throughout the first half of 1976, Appel and Haken worked on the final details of the discharging procedure, in order to produce the desired unavoidable set of reducible configurations. To do so, they sought "problem configurations" that would necessitate further changes to the discharging procedure and, on finding one, would immediately test it for reducibility—this could usually be done fairly quickly. In this way, the reducibility testing by computer could keep pace with the manual construction of the discharging procedure. The final process involved 487 discharging rules, requiring the investigation by hand of about ten thousand neighborhoods of regions with positive charge, and the reducibility testing of some two thousand configurations by computer.

Because awkward configurations sometimes took a long time to check, Appel and Haken imposed a limit of thirty minutes on each one. If it could not be proved reducible during this time, it was abandoned and replaced by other configurations: finding these was usually a straightforward process. By way of comparison, they estimated that checking the computer output for a difficult configuration by hand might take someone who was working 40-hour weeks about five years to complete.

The last few months were extremely heavy on computer time, but here, Appel, Haken, and Koch were fortunate, as the University of Illinois's computer center allowed them access when the computer was not otherwise in use. In March 1976, a powerful new computer was bought by the university's administrators, and because Appel seemed to be the only scientist who could get the machine to run properly, he initially became almost its only user, with a valuable fifty hours of computer time over the spring break.

The new computer proved to be so powerful that everything proceeded far more quickly than they had expected, saving Appel and Haken (by their own estimation) a full two years on the reducibility testing. Suddenly, by late June, and almost before they realized what was

```
N=  13 M=   9
VERTEX  1 NEIGHBORS  2  3  4  8 12 -1  0  0  0  0  0  0  0  0  0  0
VERTEX  2 NEIGHBORS 22  5  3  1 -1  0  0  0  0  0  0  0  0  0  0  0
VERTEX  3 NEIGHBORS  4  1  2  5  6  0  0  0  0  0  0  0  0  0  0  0
VERTEX  4 NEIGHBORS  3  6  7  8  1  0  0  0  0  0  0  0  0  0  0  0
VERTEX  5 NEIGHBORS  3  2 22 21  9  6  0  0  0  0  0  0  0  0  0  0
VERTEX  6 NEIGHBORS  4  3  5  9 19 18 11 10  7  0  0  0  0  0  0  0
VERTEX  7 NEIGHBORS  4  6 10 15 14 13  8  0  0  0  0  0  0  0  0  0
VERTEX  8 NEIGHBORS  4  7 13 12  1  0  0  0  0  0  0  0  0  0  0  0
VERTEX  9 NEIGHBORS  6  5 21 20 19  0  0  0  0  0  0  0  0  0  0  0
VERTEX 10 NEIGHBORS  6 11 16 15  7  0  0  0  0  0  0  0  0  0  0  0
VERTEX 11 NEIGHBORS  6 18 17 16 10  0  0  0  0  0  0  0  0  0  0  0
VERTEX 12 NEIGHBORS  1  8 13 -1  0  0  0  0  0  0  0  0  0  0  0  0
VERTEX 13 NEIGHBORS 12  8  7 14 -1  0  0  0  0  0  0  0  0  0  0  0
VERTEX 14 NEIGHBORS 13  7 15 -1  0  0  0  0  0  0  0  0  0  0  0  0
VERTEX 15 NEIGHBORS 14  7 10 16 -1  0  0  0  0  0  0  0  0  0  0  0
VERTEX 16 NEIGHBORS 15 10 11 17 -1  0  0  0  0  0  0  0  0  0  0  0
VERTEX 17 NEIGHBORS 16 11 18 -1  0  0  0  0  0  0  0  0  0  0  0  0
VERTEX 18 NEIGHBORS 17 11  6 19 -1  0  0  0  0  0  0  0  0  0  0  0
VERTEX 19 NEIGHBORS 18  6  9 20 -1  0  0  0  0  0  0  0  0  0  0  0
VERTEX 20 NEIGHBORS 19  9 21 -1  0  0  0  0  0  0  0  0  0  0  0  0
VERTEX 21 NEIGHBORS 20  9  5 22 -1  0  0  0  0  0  0  0  0  0  0  0
VERTEX 22 NEIGHBORS 21  5  2 -1  0  0  0  0  0  0  0  0  0  0  0  0

NUMBER INITIALLY GOOD   8059

PASS  1 MADE GOOD    8460  TOTAL   16519 NOT YET MADE GOOD   49911
PASS  2 MADE GOOD    9081  TOTAL   25600 NOT YET MADE GOOD   40830
PASS  3 MADE GOOD   15270  TOTAL   40870 NOT YET MADE GOOD   25560
```

#1156
35-4

The computer output for one of Appel and Haken's reducible configurations.

happening, the entire job was finished: working with his daughter, Haken had completed the construction of the unavoidable set, and within two days Appel was then able to test the final configurations for reducibility. Appel duly celebrated their achievement by placing a notice on the department's blackboard:

This phrase, *four colors suffice*, subsequently became the department's postal meter slogan.

All that remained to be done was the final checking, which needed to be carried out speedily because Appel had arranged a sabbatical visit to France and was due to leave in late July. Although they had not fully realized it, time was indeed of the essence, as they feared that several other map colorers were near to solving the four color problem—as was indeed the case:

> At the University of Waterloo in Ontario, Frank Allaire had the best reducibility methods around. By 1976, he was several months ahead of Appel and Haken in his investigations into reducibility, and was expecting to complete his solution within a few months.
>
> At the University of Rhodesia (now the University of Zimbabwe), Ted Swart, a former chemist who had carried out the first radiocarbon dating in Africa, had submitted a paper to the *Journal of*

> *Combinatorial Theory* and was informed by its editor, Bill Tutte, that Allaire was working along similar lines. Allaire and Swart pooled their results and submitted a paper just before Appel and Haken's proof was announced: this described an algorithm for determining reducibility, and listed all the reducible configurations with ring-size 10 or less.
>
> At Harvard, doctoral student Walter Stromquist had been developing powerful new methods for tackling the problem and expected to complete his solution within a year.

Although no one was aware just how close others were to completing their solutions, Appel and Haken suspected that it would be too risky to delay, especially if a rumor were to leak out that they had almost reached their goal.

They had no time to lose. Drafting in their children to help, they immediately set to work, and within a few weeks they had completed the task, producing an unavoidable set of 1936 reducible configurations. By this stage, they knew that they were safe: even if a few configurations had turned out not to be reducible after all, there was more than enough self-correction in the system for these to be replaced easily and quickly. It was not possible for a single faulty configuration, if one existed, to destroy the entire edifice.

Armed with this confidence, Appel and Haken went public, and on July 22, 1976, they formally informed their colleagues and sent out complete preprints to everyone in the field. One recipient was Bill Tutte, who two years earlier had written an article in the *American Scientist*, claiming that people who used Appel and Haken's approach were real optimists, because their method seemed extremely unlikely to work.[50] But when he heard the news, he waxed eloquent, comparing their achievement with the slaying of a fabled Norwegian sea monster:[51]

> Wolfgang Haken
> Smote the Kraken
> One! Two! Three! Four!
> Quoth he: 'The monster is no more'.

And when Tutte was interviewed by *The New York Times*, he replied:

> If they say they've done it, I have no doubt that they've done it.

Appel and Haken were delighted that a mathematician of Tutte's stature should have given his positive support so quickly. His endorsement would go a long way to setting people's minds at rest, while a lukewarm response might have cast doubts on their solution.

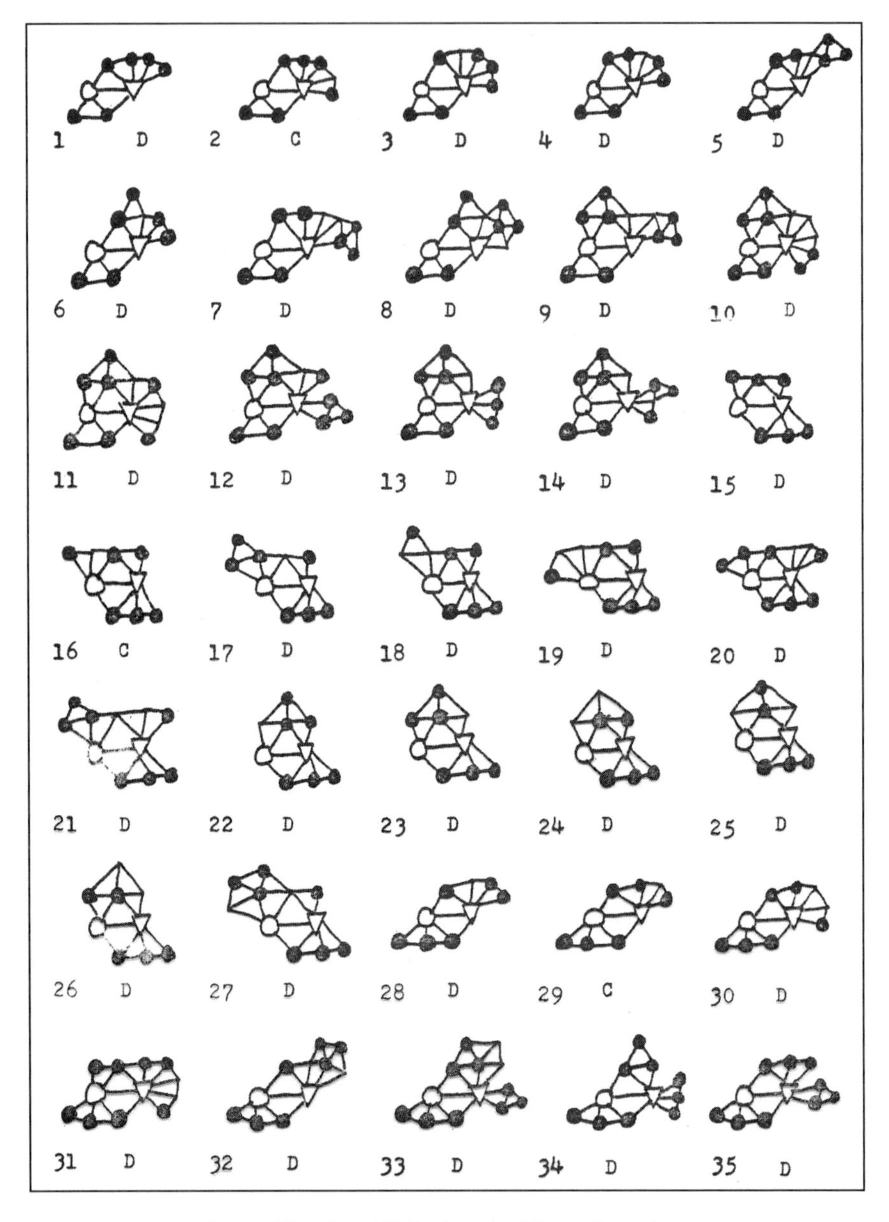

Some of Appel and Haken's reducible configurations.

On July 23, 1976, the following report appeared in *The Times* of London:[52]

> Two American mathematicians have just announced that they have solved a proposition that has been puzzling their kind for more than 100 years . . . Their proof, published today, runs to 100 pages of summary, 100 pages of

> detail and a further 700 pages of back-up work. It took each of them about 40 hours research a week and 1000 hours of computer time. Their proof contains 10,000 diagrams, and the computer printout stands four feet high on the floor.

Other newspapers around the world, from the German *Neue Zürcher Zeitung* to the Japanese *Asahi Shinbun*, latched on to the story and there was great excitement. The solution was featured in *Time*, and some of the configurations appeared on the front cover of *New Scientist*. In the September 1976 issue of its *Bulletin*, the American Mathematical Society published the following two-page "Research Announcement" by Appel and Haken, outlining the main ideas of their proof.[53] They also wrote another short note in *Discrete Mathematics* and a lengthy article in *Scientific American*.[54]

K. Appel and W. Haken: Every planar map is four colorable (1976)

Appel and Haken's research announcement opened with the following words:

> The following theorem is proved.
>
> THEOREM. *Every planar map can be colored with at most four colors.*

After remarking that they would be considering the problem in its dual formulation of coloring the vertices of a planar graph, the authors further restricted their attention to triangulations with all vertices of degree 5 or more, corresponding (in the map formulation) to cubic maps containing no digons, triangles, or squares.

They then introduced a *configuration* as a subgraph of a planar triangulation consisting of a circuit (or *ring*) and its interior, and a *reducible configuration* as one for which "it can be shown by certain standard methods that it cannot be immersed in a minimal counterexample to the four color conjecture". This referred to earlier work on reducibility by Kempe, Birkhoff, and Heesch. After recalling that "A set of configurations is called *unavoidable* if every planar triangulation contains some member of the set", they confirmed that

> It is immediate that the four color theorem is proved if an unavoidable set of reducible configurations is provided.

Appel and Haken asserted that "the most efficient known method for producing unavoidable sets" is the method of discharging, which "treats the planar triangulation as an electrical network with charge assigned to the vertices". Declaring that "the major effort in the work was involved in the development of the discharging procedure", they remarked that—although they had made extensive use of the computer when developing the discharging algorithms—the eventual algorithm was technically simpler than in their earlier approaches and had been implemented by hand. As they explained:

> The method actually produces a class of discharging algorithms which differ from one another only in minor details . . . The particular procedure chosen was determined principally to avoid configurations of ring size greater than fourteen and to employ configurations whose reducibility could be proved without exorbitant use of computer time. The algorithm produced a set U of fewer than 2000 configurations, each of ring size fourteen or smaller.

They then turned to reducible configurations, giving credit to "H. Heesch, S. Gill, and F. Allaire and E. R. Swart", and confirming the involvement of John Koch. They emphasized that their reducibility techniques "were designed for speed and efficiency in treating those configurations they could prove reducible".

Finally, they accepted that their configurations did not constitute a smallest possible such list, suggesting that minor changes in both the discharging algorithm and the reduction procedures might reduce the list by at least 25 percent. They concluded by insisting that

> It seems unlikely, however, that the theorem could be proved by these methods in a way which would avoid massive computations which required the use of computers. This last conclusion is supported by work of E. F. Moore and probabilistic calculations of the authors which indicate that such an argument always requires configurations of ring size fourteen.

Appel and Haken decided to submit their full solution to the *Illinois Journal of Mathematics*, and the resulting paper was a substantial improvement on the rough-and-ready preprint that they had sent out in July 1976. In particular, they had discovered that their preprint included repeated configurations and many instances of one configuration containing another. By eliminating these superfluous configurations, they

EVERY PLANAR MAP IS FOUR COLORABLE
PART I: DISCHARGING[1]

BY

K. APPEL AND W. HAKEN

1. Introduction

We begin by describing, in chronological order, the earlier results which led to the work of this paper. The proof of the Four Color Theorem requires the results of Sections 2 and 3 of this paper and the reducibility results of Part II. Sections 4 and 5 will be devoted to an attempt to explain the difficulties of the Four Color Problem and the unusual nature of the proof.

The first published attempt to prove the Four Color Theorem was made by A. B. Kempe [19] in 1879. Kempe proved that the problem can be restricted to the consideration of "normal planar maps" in which all faces are simply connected polygons, precisely three of which meet at each node. For such maps, he derived from Euler's formula, the equation

$$4p_2 + 3p_3 + 2p_4 + p_5 = \sum_{k=7}^{k_{max}} (k - 6)p_k + 12 \tag{1.1}$$

where p_i is the number of polygons with precisely i neighbors and k_{max} is the largest value of i which occurs in the map. This equation immediately implies that every normal planar map contains polygons with fewer than six neighbors.

In order to prove the Four Color Theorem by induction on the number p of polygons in the map ($p = \sum p_i$), Kempe assumed that every normal planar map with $p \leq r$ is four colorable and considered a normal planar map M_{r+1} with $r + 1$ polygons. He distinguished the four cases that M_{r+1} contained a polygon P_2 with two neighbors, or a triangle P_3, or a quadrilateral P_4, or a pentagon P_5; at least one of these cases must apply by (1.1). In each case he

Appel and Haken's first paper for the *Illinois Journal of Mathematics*.

were able to reduce their original list of 1936 reducible configurations to the 1482 configurations of the published version.

Their solution appeared in two parts, in the December 1977 issue of the *Illinois Journal of Mathematics*.[55] Part I, *Discharging*, written by the two of them, outlined the overall strategy of their proof and described their methods of discharging for constructing the unavoidable set. Part II, *Reducibility*, written with John Koch, described the computer implementation and listed the entire unavoidable set of reducible configurations. These were supplemented by a microfiche that contained 450 pages of further diagrams and detailed explanations.

Appel and Haken had achieved their goal:

the four color theorem was proved.

Aftermath

As we have seen, the 1960s and 1970s witnessed greatly increased activity in graph theory and combinatorics in America and around the world, but the observations that we made in Chapter 6 represent only part of the story. Alongside the opening of the University of Waterloo's Department of Combinatorics and Optimization in 1967, departments of mathematics and computer science in universities and colleges throughout America were increasingly hiring faculty members whose research interests lay in combinatorics and graph theory, and the curriculum was gradually expanded to include these and related subjects.

Meanwhile, international conferences on combinatorics and graph theory were becoming more frequent occurrences. The first of these had been a conference on combinatorics that took place in 1959 in Dobogókő, Hungary, and this was followed by other European meetings in Germany, Czechoslovakia, Hungary, and England. In the United States, the Chapel Hill combinatorics conferences at the University of North Carolina commenced in 1967, to be followed a year later by the first of the Kalamazoo quadrennial meetings on graph theory at Western Michigan University. The series of Southeastern International Conferences on Combinatorics, Graph Theory, and Computing began in 1970 at Louisiana State University in Baton Rouge, and these annual meetings have continued to this day, now taking place at Florida Atlantic University in Boca Raton.

Several of these meetings issued conference proceedings, and there was a corresponding increase in other books on graph theory and combinatorics. Those by Claude Berge and Oystein Ore, mentioned in Chapter 6, were succeeded by Robert G. Busacker and Thomas L. Saaty's *Finite Graphs and Networks: An Introduction with Applications*, which appeared in 1965, and these were soon complemented by Mehdi Behzad and Gary Chartrand's *Introduction to the Theory of Graphs* in

1971, Robin Wilson's *Introduction to Graph Theory*, in 1972, and J. A. Bondy and U.S.R. Murty's *Graph Theory with Applications* in 1976.[1] In combinatorics, there were corresponding introductions to the subject by John Riordan, Herbert Ryser, Marshall Hall Jr., C. L. (Dave) Liu, and Ian Anderson.[2]

Meanwhile, books on specific areas of graph theory began to appear with increasing frequency, to join Ford and Fulkerson's *Flows in Networks*, Harary, Norman, and Cartwright's *Structural Models*, and Harary and Palmer's *Graphical Enumeration* (mentioned in Chapter 5), and Ore's *The Four-Color Problem*, Ringel's *Map Color Theorem*, and Garey and Johnson's *Computers and Intractability* (see Chapter 6). These new works included two volumes edited by Frank Harary, *Graph Theory and Theoretical Physics* and *A Seminar on Graph Theory* (1967), and monographs by W. T. Tutte, *Connectivity in Graphs* (1966), and John Moon, *Topics on Tournaments* (1968). Among those published in the 1970s were books by John Moon, *Counting Labelled Trees* (1970), Richard Bellman, Kenneth L. Cooke, and Jo Ann Lockett, *Algorithms, Graphs and Computers* (1970), Arthur T. White, *Graphs, Groups and Surfaces* (1973), Paul Erdős and Joel Spencer, *Probabilistic Methods in Combinatorics* (1974), and Fred S. Roberts, *Discrete Mathematical Models* (1976).[3] Also in 1976, N. L. Biggs, E. K. Lloyd, and R. J. Wilson produced *Graph Theory 1736–1936*, on the early history of the subject.[4] A couple of years later, Lowell W. Beineke and Robin J. Wilson edited *Selected Topics in Graph Theory* and *Applications of Graph Theory*,[5] the first of their many collections of survey chapters by well-known experts.

New journals emphasizing research in graph theory were also being developed. In 1966, the *Journal of Combinatorial Theory* was founded by Frank Harary and Gian-Carlo Rota, later to be divided into Series A on combinatorial structures and Series B on graphs. This was followed soon after by *Discrete Mathematics* in 1971, the *Journal of Graph Theory* in 1977, and many later ones.

SOME RESEARCH TOPICS

On the research side, there was also an explosion of activity in the 1960s and 1970s. Alongside the subjects covered in Chapter 6, new areas of study within graph theory were developing, including generalized Ramsey theory, generalized colorings, traversability, connectivity, random graph theory, extremal graph theory, graph decompositions, labelings, and coverings. The following topics are among those explored at this time by graph theorists from the United States and Canada.

Moore Graphs

The *diameter* of a graph is the maximum distance between any two vertices of the graph, and the *girth* is the length of the shortest cycle in the graph. A *Moore graph* (named after Edward F. Moore) is a regular graph with diameter k and girth $2k+1$. In 1960, Alan Hoffman and Robert Singleton of IBM proved that the only Moore graphs with diameter 2 are graphs that are regular of degree 2 (the cycle with five vertices), degree 3 (the Petersen graph), degree 7 (the so-called "Hoffman–Singleton graph"), and possibly one of degree 57 whose existence is still unknown.[6] The only Moore graph with diameter 3 is the cycle with seven vertices, and there are no Moore graphs with diameter greater than 3.

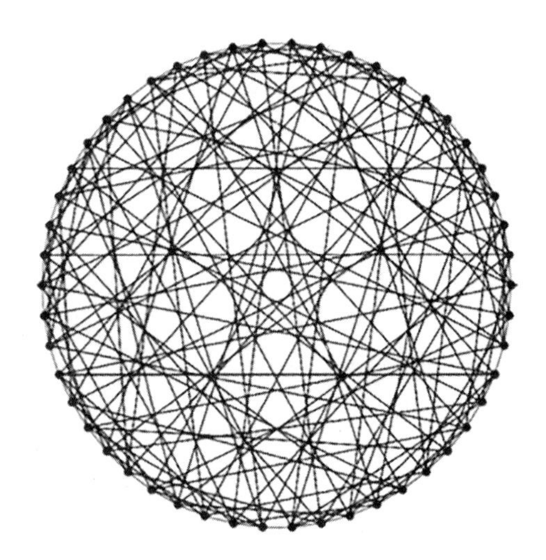

The Hoffman–Singleton graph with 50 vertices.

Degree Sequences

The *degree sequence* of a graph is a list of the degrees of the vertices, usually in decreasing order. In 1962, the Iranian–American graph theorist S. L. Hakimi, of Northwestern University in Illinois, characterized those lists that are the degree sequences of graphs.[7] His well-known result had earlier been discovered independently by Václav Havel and published in Czech.

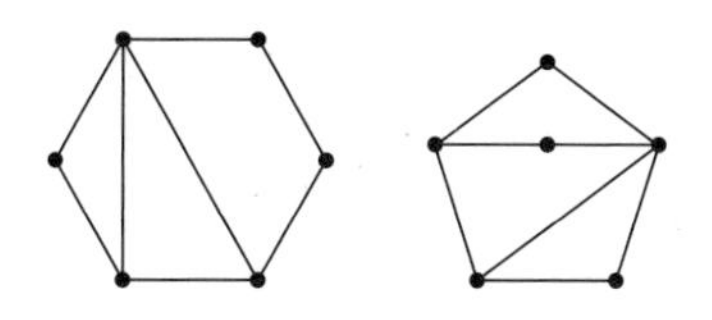

Two graphs with degree sequence (4, 3, 3, 2, 2, 2).

Tournaments

A *tournament* is a complete graph in which a direction has been assigned to each edge. In 1964, John Moon of the University of Alberta in Edmonton, Canada, explained why no tournament has a group of symmetries with an even number of elements. He also showed that every abstract group with an odd number of elements is the group of symmetries of a tournament—in fact, of infinitely many.[8]

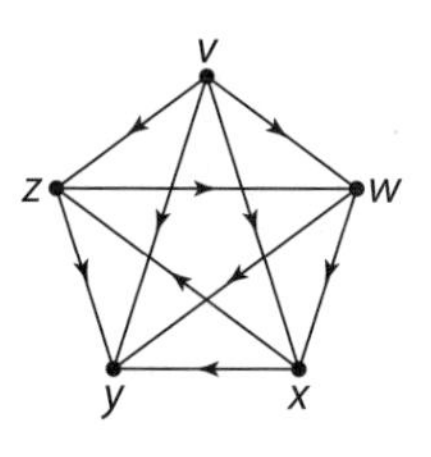

A tournament with five vertices.

Matroids

Following Bill Tutte's characterizations of graphic and cographic matroids in 1959 (see Chapter 5), his interest in the subject continued. In 1964, the National Bureau of Standards in Washington, DC, hosted the first conference on matroid theory, organized by Jack Edmonds, and at this meeting Tutte presented some notable "Lectures on matroids". These were then published in the conference's proceedings, together with an analog for matroids of Menger's theorem that had originally appeared in Tutte's Cambridge doctoral thesis.[9]

Line Graphs

An *induced subgraph* of a graph G is a subgraph H with the property that every edge of G with both of its ends in H is also an edge of H. As we saw in Chapter 4, the *line graph* $L(G)$ of a graph G is the graph whose vertices correspond to the edges of G, with two vertices adjacent whenever the corresponding edges of G meet at a common vertex. In 1968,

Lowell W. Beineke of Purdue University, Fort Wayne, Indiana, characterized line graphs in terms of nine forbidden induced subgraphs.[10]

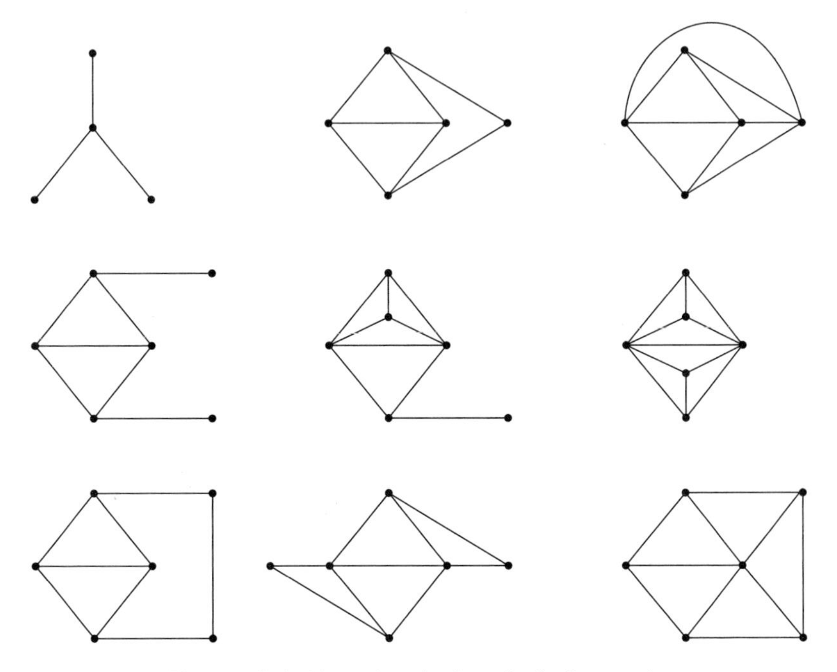

The nine forbidden induced subgraphs for line graphs.

Probability

In his early papers on Ramsey numbers, Paul Erdős introduced the *probabilistic method* used to prove the existence of certain mathematical objects with a particular property (such as being a certain type of graph) by showing that the probability of an object existing with that property is strictly positive. Such proofs give no clue as to how to construct a specific object of this type. The first construction of this kind seems to have been produced by Ron Graham and Joel Spencer in 1971, for a specific type of tournament whose existence had been predicted eight years earlier by Erdős using the probabilistic method.[11]

Hamiltonian Graphs

In Chapter 6, we presented Oystein Ore's theorem, that a graph with n vertices is Hamiltonian if $\deg(v) + \deg(w) \geq n$ whenever the vertices v and w are not adjacent. This result was generalized in 1971 by J. Adrian Bondy of the University of Waterloo.[12] He defined a graph G with n vertices to be *pancyclic* if it contains cycles of all lengths from 3 to n, and proved that if Ore's condition holds, then the graph is either pancyclic or is the complete bipartite graph $K_{n/2,n/2}$. Later, he developed these ideas with Václav Chvátal to prove that a graph is Hamiltonian if and only if its closure is Hamiltonian, where the *closure* of a graph is the graph (not necessarily complete) obtained by successively joining pairs of non-adjacent vertices whose degrees add up to at least n.

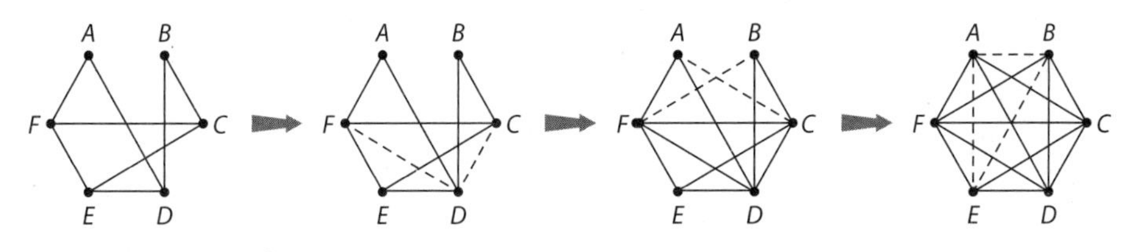

Forming the closure of a graph.

Voltage Graphs

Also in Chapter 6, we saw how William Gustin introduced current graphs, where Kirchhoff's current law is satisfied, when solving the "thread problem" of drawing complete graphs on orientable surfaces. In 1974, Jonathan Gross of Columbia University introduced the dual idea of a *voltage graph*, where Kirchhoff's voltage law is satisfied.[13] These have turned out to be a more natural and flexible setting for some problems on the embedding of graphs on surfaces.

Graph Embeddings

In Interlude B, we met Kuratowski's theorem, which states that a graph is planar if and only if it contains no subgraph that is homeomorphic to the complete graph K_5 or the complete bipartite graph $K_{3,3}$. The question then arises whether there are corresponding lists of "forbidden subgraphs" for other surfaces. In 1977, Henry Glover and John P. Huneke of the Ohio State University proved that a graph can be embedded in a projective plane if and only if it contains no subgraph that is homeomorphic to one of 103 graphs.[14] This was the first surface, other than the sphere, for which an explicit list of forbidden subgraphs had been produced.

THE GRAPH MINORS PROJECT

Central to the subsequent development of graph theory has been the "Graph Minors Project", an extended program of activity undertaken by Neil Robertson of Ohio State University and Paul Seymour, originally from Oxford University, but later resident at Ohio State before moving to Princeton University. Working together with various collaborators for two months of every summer from around 1983 to 2004, they succeeded in solving an astonishing number of difficult problems in graph theory, including the following:

Generalizing Kuratowski's Theorem

Continuing on from the results of Kuratowski, and of Glover and Huneke (as mentioned above), Robertson and Seymour announced in 1984 that they had proved that there is a finite collection of forbidden subgraphs for any surface, whether orientable or non-orientable. Their full proof of this remarkable result was eventually published in 2004.[15]

Hadwiger's Conjecture

In 1943, the Swiss mathematician Hugo Hadwiger conjectured that every connected k-colorable graph could be reduced to the complete graph K_k by contracting edges (see Chapter 5).[16] Hadwiger proved this conjecture for $k=2$, 3, and 4, and its truth for $k=5$ was shown to be equivalent to the four color theorem, and was thereby confirmed in 1976. In 1993, Robertson and Seymour, together with Robin Thomas, proved Hadwiger's conjecture for $k=6$, after a long and difficult struggle that again used the four color theorem.[17] It is not known whether the conjecture is true when $k>6$.

The Four Color Theorem

Appel and Haken's somewhat ad hoc proof of the four color theorem was greeted with enthusiasm, but also with dismay by those who questioned the validity of a computer-assisted proof whose details could not be checked by hand. Around 1994, by using essentially the same approach as Appel and Haken had, but applying it more systematically, Robertson and Seymour, in collaboration with Daniel Sanders and Robin Thomas, presented a new proof which involved only 633 reducible configurations and could be checked directly on one's own personal computer in a matter of hours.[18] This proof was widely accepted and later formally verified as correct.[19]

Perfect Graphs

In 1961, Claude Berge defined a graph to be *perfect* if the chromatic number and the clique number (the size of the largest complete subgraph) are equal for the graph and for each of its induced subgraphs, and he made the following conjecture:[20]

> *Strong perfect graph conjecture*: A graph is perfect if and only if it contains no odd cycle or its complement as an induced subgraph.

In 1972, László Lovász had proved the following weaker form of this conjecture:[21]

> A graph is perfect if and only if its complement is perfect.

The strong form of the conjecture was much more difficult to prove. This was eventually achieved in 2002 by Robertson and Seymour, together with Maria Chudnovsky and Robin Thomas, and published in 2006.[22]

* * * * *

The development of graph theory in America over the century from 1876 to 1976 was truly remarkable—from James Joseph Sylvester's appointment at Johns Hopkins University, and the later advances by Oswald Veblen, George Birkhoff, Philip Franklin, and Hassler Whitney, to the subsequent development of graph algorithms, the achievements of Bill Tutte, and the proofs of the Heawood conjecture by Gerhard Ringel and Ted Youngs, and the four color theorem by Kenneth Appel and Wolfgang Haken. These were not the only mathematicians in North America to advance the subject, as we have seen, but they were undoubtedly among the most significant, as they helped to prepare the ground for the subsequent explosion of activity by many thousands of graph theorists throughout the world.

GLOSSARY

abstract dual A form of graph duality in which two connected graphs G and G^* are duals if there is a one–one correspondence between their edge-sets with the property that a set of edges in G forms a cycle of G if and only if the corresponding set of edges of G^* forms a cutset of G^*; this form of duality was introduced by Hassler Whitney in 1933.

adjacent edges Two edges in a graph that share a common vertex.

adjacent vertices Two vertices in a graph that are joined by an edge.

algorithm A finite step-by-step process for solving a problem.

analysis situs The study of position or situation, now called *topology*.

assignment problem A matching problem in a weighted bipartite graph where the object is to assign applicants in an optimal way to jobs for which they are qualified.

balanced signed graph A signed graph in which every cycle has an even number of negative edges.

base of a matroid A maximal independent set in a matroid; for a connected graph, the bases are the spanning trees.

binary matroid A matroid that can be represented as a set of vectors in a vector space over the field of two elements, 0 and 1.

bipartite graph A graph whose vertices can be divided into two sets A and B so that every edge joins a vertex in A to a vertex in B.

Birkhoff diamond A reducible configuration consisting of four pentagons surrounded by a ring of six regions; it was introduced by G. D. Birkhoff in 1913.

block A maximal connected subgraph with no cut-vertices—that is, a maximal non-separable graph.

boundary An edge bordering two neighboring countries (regions) of a map.

Brahana's theorem Every surface is topologically equivalent to either some orientable surface or some non-orientable surface.

breadth-first search A search method for graphs in which one starts at an arbitrary vertex and explores all of its neighboring vertices before moving on to more distant vertices.

capacitated network A directed graph in which each arc is assigned a positive number (its capacity) that represents the maximum amount of flow allowed along it.

Cayley's theorem The number of labeled trees on n vertices, or of spanning trees in the complete graph K_n, is n^{n-2}.

ceiling function The function that rounds up a real number x and whose output is denoted by $\lceil x \rceil$; for example, $\lceil \pi \rceil = 4$, $\lceil -\pi \rceil = -3$, and $\lceil 7 \rceil = 7$.

chromatic number of a graph The smallest number k for which the vertices of a graph can be colored with k colors, so that adjacent vertices are colored differently.

chromatic number of a map The smallest number k for which the regions of a map can be colored with k colors, so that neighboring regions are colored differently.

chromatic number of a surface The smallest number k for which the vertices of any graph, or the regions of any map, drawn on the surface can be colored with k colors with adjacent vertices, or neighboring regions, colored differently.

chromatic polynomial of a graph The number of ways of coloring the vertices of a graph with a given number of colors, so that adjacent vertices receive different colors; it is always a polynomial in the number of colors.

chromatic polynomial of a map The number of ways of coloring the regions of a map with a given number of colors, so that neighboring regions receive different colors.

circuit of a matroid A minimal dependent set in a matroid; for a graph, the circuits are the cycles.

cographic matroid A matroid that is the cutset matroid of some graph.

coloring of a graph or map An assignment of colors to the vertices of a graph, or the regions of a map, so that adjacent vertices, or neighboring regions, are colored differently.

combinatorial dual A form of dual graph whose definition depends on the rank and nullity of its subgraphs; introduced by Hassler Whitney, it is sometimes called the *Whitney dual*.

combinatorial topology An earlier name for what is now called *algebraic topology*.

combinatorics The branch of mathematics that is primarily concerned with the arrangement and enumeration of elements of a (usually finite) set.

complete bipartite graph A bipartite graph whose vertices are divided into two sets A and B such that every vertex in A is joined to every vertex in B by an edge; if the sets A and B have r and s vertices, the complete bipartite graph is denoted by $K_{r,s}$.

complete graph A graph in which every vertex is joined to every other vertex; the complete graph with n vertices is denoted by K_n.

complete matching See perfect matching.

complete set of cycles A set of cycles in a graph with the property that every cycle of the graph can be written as a sum (modulo 2) of cycles in the set.

complexity The study of the efficiency of algorithms for solving problems.

component The separate pieces of a graph; in each component there is at least one path between each pair of vertices.

configuration in a graph A collection of vertices that are surrounded by an outside ring of vertices.

configuration in a map A collection of regions that are surrounded by an outside ring of regions.

connected component See component.

connected graph A graph that is in one piece, so that there is at least one path between each pair of vertices.

connectivity of a graph A measure of how connected a connected graph is; a connected graph is *k-connected* if k is the smallest number of vertices whose deletion disconnects the graph or leaves a single vertex.

contracting an edge Removing an edge vw and identifying the vertices v and w so that all edges that were formerly incident with either v or w are now incident to the new vertex.

Counting theorem If, for each $k \geq 2$, C_k is the number of k-sided regions in a cubic map, then

$$4C_2 + 3C_3 + 2C_4 + C_5 - C_7 - 2C_8 - 3C_9 - 4C_{10} - \cdots = 12.$$

***C*-reducible configuration** A configuration that can be proved to be reducible only after it has been modified in some way.

critical path analysis The study of problems involving the scheduling of key tasks in a project; it usually involves finding longest paths between the vertices of a graph or network.

cross-cap To add a cross-cap to a surface, cut a hole in the surface, and identify the boundary of the hole with the edge of a Möbius band.

cubic graph A graph in which exactly three edges meet at each vertex

cubic map A map in which exactly three regions and three boundary lines meet at each point.

current graph A network of arcs labeled with numbers that satisfy Kirchhoff's current law; current graphs were used in the proof of the Heawood conjecture to produce rotation schemes, and in solving the problem of squaring the square.

cutset (or **minimal cutset**) A set of edges whose removal splits a connected graph into two pieces, and is minimal with respect to this property.

cutset matroid A matroid whose circuits correspond to the cutsets of a graph.

cut-vertex A vertex whose deletion increases the number of components of a graph.

cycle A sequence of distinct alternating vertices and edges of a graph (where each edge joins the preceding and succeeding vertices) that returns to the starting vertex; a cycle can be represented by a list of the form

$$v_0, e_1, v_1, e_2, v_2, \ldots, v_{n-1}, e_n, v_0.$$

cycle matroid A matroid whose circuits correspond to the cycles of a graph.

degree of a vertex The number of edge-ends attached to that vertex.

deleting an edge Removing an edge vw from a graph, leaving the vertices v and w.

deleting a vertex Removing from a graph a vertex and all of its incident edges.

dependent set in a matroid A minimal set of elements that contains a circuit.

depth-first search A search method for graphs in which one starts at an arbitrary vertex and penetrates the graph as deeply as possible before backtracking to other vertices.

diameter of a graph The maximum distance between any two vertices of the graph.

digon A two-sided region in a map.

discharging See method of discharging.

distance between two vertices The number of edges in a shortest path joining the vertices.

***D*-reducible configuration** A configuration in which every coloring of the regions in the surrounding ring is a proper coloring, or may be converted into one by applying the method of Kempe chains.

dual graph To construct a (geometric) dual G^* of a plane drawing of a planar graph G, place a vertex of G^* inside each region of G (including the external region), and for each edge of G join the vertices of G^* in the regions on each side of the edge with an edge of G^* crossing the edge of G.

dual matroid If M is a matroid with bases B, then its dual matroid is the matroid whose bases are the complements of the bases of M.

edge A line or curve joining two vertices of a graph or a boundary line between two regions of a map.

Euler's formula (for polyhedra) If a polyhedron has F faces, E edges, and V vertices, then $F-E+V=2$; this was first observed by Leonhard Euler in 1750.

Euler's formula for maps on a plane or sphere For any map drawn on a plane or sphere,

$$(\text{number of regions}) - (\text{number of edges}) + (\text{number of vertices}) = 2.$$

Euler's formula for maps on a non-orientable surface For any map drawn on the non-orientable surface N_q with q cross-caps,

$$(\text{number of regions}) - (\text{number of edges}) + (\text{number of vertices}) = 2-q.$$

Euler's formula for maps on an orientable surface For any map drawn on the orientable surface S_g with g handles,

$$(\text{number of regions}) - (\text{number of edges}) + (\text{number of vertices}) = 2-2g.$$

exponential-time algorithm An algorithm whose running time is proportional to k^n (for some number $k>1$), where n is the input size.

external region The unbounded region of a map or a plane drawing of a planar graph.

Fano matroid A matroid whose bases are all triples of integers from 1 to 7, except those corresponding to the lines of a Fano plane.

Fano plane A geometric configuration with seven points and seven lines, with three points on each line, and three lines through each point.

floor function The function that rounds down a real number x and whose output is denoted by $\lfloor x \rfloor$; for example, $\lfloor \pi \rfloor = 3$, $\lfloor -\pi \rfloor = -4$, and $\lfloor 7 \rfloor = 7$.

flow in a network An allocation of a non-negative number to each edge of a capacitated network so that the flow along each arc does not exceed the capacity of the arc and the total flow into any vertex (other than the start and terminal vertices) equals the total flow out of it.

four color problem for graphs Can the vertices of every planar graph be colored with at most four colors so that adjacent vertices are colored differently?

four color problem for maps Can the regions of every map drawn on a plane or the surface of a sphere be colored with at most four colors so that neighboring regions are colored differently?

four color theorem The regions of every map drawn on a plane or the surface of a sphere can be colored with at most four colors so that neighboring regions are colored differently. Equivalently, the vertices of every planar graph can be colored with at most four colors so that adjacent vertices are colored differently.

genus of a surface An orientable surface is of genus g if it is topologically equivalent to a sphere with g handles; examples are the sphere ($g=0$) and the torus ($g=1$).

A non-orientable surface is of genus q if it is topologically equivalent to a sphere with q cross-caps; examples are the projective plane ($q=1$) and the Klein bottle ($q=2$).

geographically good configuration A configuration that contains neither of Heesch's first two obstacles to reducibility.

geometric dual See dual graph.

girth of a graph The number of edges in a shortest cycle of the graph.

good algorithm Another name for a polynomial-time algorithm.

good coloring A coloring of a ring of regions that can be extended directly to a coloring of the regions within the ring.

graph A set of vertices (usually finite) and a set of unordered pairs of vertices, called edges, where each edge joins two vertices; when an edge joins a vertex to itself it is called a loop.

graphic matroid A matroid that is the cycle matroid of some graph.

greedy algorithm An algorithm in which one makes the optimal choice at each stage; an example is Kruskal's algorithm for the minimum spanning tree problem.

Hadwiger's conjecture Every connected k-colorable graph can be reduced to the complete graph K_k by contracting some edges.

Hall's "marriage" theorem Suppose that each of a collection of boys is acquainted with a collection of girls. Then each boy can marry one of his acquaintances if and only if, for each number k, every set of k boys is collectively acquainted with at least k girls.

Hamiltonian cycle A cycle in a graph that includes every vertex.

Hamiltonian graph A graph that has a Hamiltonian cycle.

Heawood conjecture for a non-orientable surface For each $q \geq 1$, the chromatic number of the surface N_q is $\chi(N_q) = \lfloor \frac{1}{2}(7 + \sqrt{1 + 24q}) \rfloor$, except that $\chi(N_2) = 6$; equivalently, the non-orientable genus of K_n ($n \geq 3$) is $\hat{g}(K_n) = \lceil \frac{1}{6}(n-3)(n-4) \rceil$, except that $\hat{g}(K_7) = 3$.

Heawood conjecture for an orientable surface For each $g \geq 1$, the chromatic number of the surface S_g is $\chi(S_g) = \lfloor \frac{1}{2}(7 + \sqrt{1 + 48g}) \rfloor$; equivalently, the orientable genus of K_n ($n \geq 3$) is $g(K_n) = \lceil \frac{1}{12}(n-3)(n-4) \rceil$.

Heesch's obstacles to reducibility See obstacles to reducibility.

Hoffman–Singleton graph A Moore graph with diameter 2 and girth 5. It has fifty vertices and is regular of degree 7.

homeomorphic graphs Two graphs are homeomorphic if they can be obtained from each other by the insertion or removal of vertices of degree 2.

Hungarian method An efficient algorithm for solving assignment problems; it was developed by H. W. Kuhn, based on earlier results of Dénes König and Jenö Egerváry.

incidence If $e = vw$ is an edge of a graph, then e is incident to both v and w, and v and w are both incident to e.

independent set in a matroid A set of elements that contains no circuit; for a graph, the independent sets consist of trees.

induced subgraph A subgraph H of a graph G such that every edge of G with both of its ends in H is also an edge of H.

irreducible configuration A configuration that is not reducible.

irreducible map A minimal counter-example to the four color theorem; the proof of the theorem in 1976 showed that such maps do not exist.

isomers Molecules with the same chemical formula but with different atomic configurations.

isomorphic graphs Graphs for which there are one–one correspondences between their vertex-sets that preserve the adjacency of vertices.

Kempe chain A chain of regions in a map, or a path in a graph, that is colored with two alternating colors.

Kempe-chain argument A procedure used in map or graph coloring in which two colors are interchanged along a chain of regions or vertices to enable the coloring of a region or vertex that could not previously be colored.

Kirchhoff's current laws The sum of the currents entering each vertex of an electrical network equals the sum of the currents leaving it, and the sum of the currents around any cycle is 0.

Klein bottle A non-orientable surface of genus 2 that can be constructed by gluing the top and bottom sides of a rectangle together, and then gluing the other two sides of the rectangle together but in opposite directions; this cannot be done in three dimensions without the resulting "bottle" self-intersecting. The Klein bottle is topologically equivalent to a sphere with two cross-caps and was first described by Felix Klein in 1882.

König's theorem In any bipartite graph, the maximum size of a matching is equal to the minimum number of vertices that collectively meet all the edges.

Kruskal's algorithm An algorithm that is used to solve the minimum spanning tree problem for a weighted graph by successively selecting an edge not previously chosen of smallest weight that does not create a cycle; it is a greedy algorithm that is based on a method described by O. Borůvka in 1926 and was rediscovered in 1956 by J. Kruskal.

Kuratowski's theorem A graph is planar if and only if it has no subgraph that is homeomorphic to the complete graph K_5 or the complete bipartite graph $K_{3,3}$.

leaf A vertex of degree 1; a previous meaning, used, for example, in Petersen's theorem, was a part of a graph that could be separated from the rest by the removal of a single edge.

line graph For a connected graph G, the graph $L(G)$ whose vertices correspond to the edges of G, with two vertices of $L(G)$ adjacent whenever the corresponding edges of G are adjacent.

linear programming A method to optimize the outcome in a situation represented by given linear equations and inequalities.

linkage A planar graph obtained by marking a point in each region of a map and joining the points in neighboring regions.

longest path problem The problem of determining a longest route between two vertices of a network or weighted graph; such problems arise in critical path analysis.

loop An edge joining a vertex to itself.

manifold An n-dimensional space that near each point resembles a Euclidean space; examples are a circle, sphere, torus, and the Klein bottle.

matching A set of edges in a graph (often bipartite) with no vertices in common.

matching problem The problem of finding a matching in a given graph; an example is the matching of applicants to jobs for which they are variously qualified.

matroid An abstract notion that generalizes the ideas of rank and independence in graph theory and of dimension and linear independence in vector spaces.

max-flow min-cut theorem In a capacitated network, the value of a maximum flow from the start to the terminal is equal to the capacity of a minimum cut.

meeting point in a map A point where countries (regions) and boundary lines meet.

Menger's theorem Let v and w be non-adjacent vertices in a connected graph. Then the maximum number of vertex-disjoint paths joining v and w is equal to the minimum number of vertices whose removal separates v from w.

method of discharging A procedure that involves the movement of (electrical) charges around a graph; it is used for determining whether a given set of configurations is an unavoidable set.

method of Kempe chains See Kempe-chain argument.

minimax theorem A theorem that asserts that the minimum of one quantity in a graph is equal to the maximum of another quantity; examples are König's theorem, Menger's theorem, and the max-flow min-cut theorem.

minimum connector problem See minimum spanning tree problem.

minimum spanning tree problem The problem of finding a spanning tree of minimum total length in a weighted graph; see Kruskal's algorithm.

Möbius band A one-sided object constructed from a rectangular strip by identifying its ends in opposite directions.

Moore graph A regular graph with diameter k and girth $2k+1$, for some number k; examples include the Petersen graph and the Hoffman–Singleton graph.

multiple edges More than one edge joining a pair of vertices.

neighboring regions Two regions that share a common boundary line.

non-orientable genus of a graph The smallest number q for which the graph can be drawn without crossings on the non-orientable surface N_q.

non-orientable surface A surface in which clockwise rotation is not maintained when moving on the surface; examples are a Möbius band, projective plane, Klein bottle, and a sphere with cross-caps attached.

non-separable graph A connected graph that cannot be disconnected by removing a single vertex (a cut-vertex).

NP The set of problems where solutions, once obtained, can be checked in polynomial time; NP is short for non-deterministic polynomial-time problem.

NP-complete problem A problem whose solution in polynomial time implies that every NP problem can be solved in polynomial time.

nullity of a graph $E-V+P$, where E is the number of edges, V is the number of vertices, and P is the number of components.

obstacles to reducibility Three arrangements of countries, introduced by Heinrich Heesch, whose appearance in a configuration seems to indicate that it is not reducible.

orientable genus of a graph The smallest number g for which the graph can be drawn without crossings on the orientable surface S_g.

orientable surface A surface in which the concept of clockwise rotation is maintained when moving in a continuous manner on the surface; examples are a plane, sphere, torus, and a sphere with handles attached.

P The set of problems that can be solved in polynomial time.

pancyclic graph A graph with n vertices that contains cycles of all lengths from 3 to n.

perfect graph A graph whose chromatic number and clique number (the size of its largest complete subgraph) are equal for the graph and each of its induced subgraphs.

perfect graph theorem Weak form (1972): a graph is perfect if and only if its complement is perfect. Strong form (2002): a graph is perfect if and only if it contains no odd cycle or its complement as an induced subgraph.

perfect matching (also called a complete matching or 1-factor) A collection of disjoint edges that includes every vertex of a graph.

Petersen graph A cubic graph with ten vertices and fifteen edges; it was introduced by Julius Petersen in 1898.

Petersen's theorem Every cubic graph with at most two leaves has a 1-factor.

planar graph A graph that can be drawn on a plane or sphere without any edges crossing.

planar matroid A matroid that is both graphic and cographic.

plane drawing of graph A drawing of a graph in which no two edges intersect, except at a vertex where both are incident.

polyhedron A 3-dimensional shape with flat polygonal faces, straight edges, and sharp corners (or vertices).

polynomial An algebraic expression such as $x^4 - 5x^3 + 8x^2 - 4x + 3$.

polynomial-time algorithm An algorithm whose running time is proportional to n^k (for some positive number k), where n is the input size.

projective plane A surface obtained by identifying opposite sides of a rectangle in opposite directions; it is the surface with non-orientable genus 1.

Ramsey graph theory An area of graph theory that investigates the number of vertices that a graph must have in order to ensure the appearance of subsets of a specified type.

Ramsey number The minimum number $N = r(m, n)$ of vertices needed to ensure that any red–blue coloring of the edges of K_N contains either a red K_m or a blue K_n.

rank of a graph $V - P$, where V is the number of vertices of the graph and P is the number of components.

rank of a matrix The maximum number of linearly independent columns in the matrix.

rank of a subset in a matroid The number of elements in a largest independent set in the subset.

reducible configuration A configuration that cannot occur in a minimal counter-example to the four color theorem; if a map contains a reducible configuration, then any coloring of the rest of the map with four colors can be extended (possibly after some recoloring) to a coloring of the entire map.

region of a map A general term for a country or county in a map; it may also refer to the external region.

region of a planar graph An area surrounded by edges on a plane drawing of the graph; one of these regions is unbounded and is called the external region.

regular graph A graph in which every vertex has the same degree.

regular matroid A matroid that is representable over every field.

representable matroid A matroid that can be represented as a set of vectors in a vector space over some field.

***r*-factor** A regular subgraph of degree r that includes every vertex of a graph; in particular, a 1-factor is often called a "perfect matching" or a "complete matching" and is a collection of disjoint edges that includes every vertex of the graph.

***r*-factorization** An r-factorization of a graph splits the graph into disjoint r-factors.

ring (in a configuration) The regions that bound a configuration.

ring-size The number of regions in the ring that bounds a configuration; it is a measure of the configuration's complexity.

Ringel–Youngs theorem The theorem of Gerhard Ringel and J.W.T. Youngs that proved the Heawood conjecture.

rooted tree A tree in which one particular vertex is designated as the "root".

rotation scheme In a colored map, a list that specifies for every region the colors of the regions that surround it. In a colored complete graph, a list that specifies for every vertex the colors of the vertices adjacent to it; in each case, the colors appear in counter-clockwise order.

saturated arc An arc in a capacitated network for which the flow along it equals its capacity.

search tree A tree network in which the aim is to visit every vertex in a specified manner.

separable graph A connected graph that can be disconnected by the removal of a single vertex (cut-vertex); removing all such cut-vertices splits the graph into smaller pieces called "blocks".

seven color theorem for a torus The regions of every map on a torus can be colored with at most seven colors, with neighboring regions colored differently, and there are maps that require seven colors.

Shimamoto horseshoe A configuration with ring-size 14 whose D-reducibility would have implied the four color theorem.

shortest path problem The problem of determining a shortest route between two given locations in a network or weighted graph.

signed graph A graph in which each edge is designated as either positive or negative.

simple graph A graph with no loops or multiple edges.

simply connected surface A surface which is connected (any two points can be joined by a path) and has no "holes" (any loop joining a point to itself can be shrunk to a single point); for example, a plane and a sphere are simply connected, but a torus is not.

spanning tree A tree in a connected graph which includes every vertex of the graph.

squaring the square (or rectangle) The problem of finding a square (or rectangle) with integer sides that can be divided into squares with unequal integer sides.

subgraph A graph *H* is a subgraph of *G* if all the vertices and edges of *H* appear in *G*.

surface The two types of surfaces are orientable surfaces (such as a sphere or torus) and non-orientable surfaces (such as a projective plane or Klein bottle).

system of neighboring regions A map in which every region meets every other one.

thread problem Another name for the complete graph version of the Heawood conjecture.

topology The branch of mathematics which deals with geometric properties of objects that are preserved under continuous deformations such as stretching or twisting.

torus A two-dimensional surface that looks like an inner tube or the surface of a bagel; formally, a torus is generated by rotating a circle through 360° around an axis outside the circle.

tournament A complete graph for which a direction has been assigned to each edge.

transportation problem The problem of determining how to distribute commodities from factories to markets in a network at minimum cost.

traveling salesman problem Given a number of cities connected by links that join cities in pairs, and given the distances between all pairs of cities, what is a shortest cyclic route that visits every city?

tree A connected graph with no cycles.

triangulation A plane graph or map in which each region is bounded by three edges.

Tutte polynomial A polynomial in two variables which represents many properties of graphs and has numerous applications in other areas as well.

unavoidable set of configurations A collection of configurations, at least one of which must appear in every map.

utilities problem Can one join three utilities (water, gas, and electricity) to three houses without any pipes crossing? This problem is equivalent to that of determining whether the complete bipartite graph $K_{3,3}$ can be drawn in the plane without any edges crossing.

value of a flow The total flow from the start vertex to the terminal vertex in a capacitated network.

vertex A point in a graph or map where edges meet.

Whitney dual See combinatorial dual.

NOTES, REFERENCES, AND FURTHER READING

In these notes, we use the following abbreviations for works that are referred to throughout this book:

Cayley Papers: A. Cayley, *Mathematical Papers*, 13 volumes and index volume, Cambridge University Press, 1889–97.
Century of Mathematics: *A Century of Mathematics in America*, Parts I, II, III (ed. Peter Duren, assisted by Richard A. Askey and Uta C. Merzbach), American Mathematical Society (Providence), 1989.
Four Colors Suffice: Robin Wilson, *Four Colors Suffice: How the Map Problem Was Solved*, Princeton University Press, 2014.
Graph Theory 1736–1936: N. L. Biggs, E. K. Lloyd, and R. J. Wilson, *Graph Theory 1736–1936*, Clarendon Press (Oxford), 1976; later editions 1986, 1998.

Further biographical information about many of the people discussed in this book can be found in the MacTutor History of Mathematics Archive of the University of St Andrews, Scotland, *https://mathshistory.st-andrews.ac.uk*.

SETTING THE SCENE: EARLY AMERICAN MATHEMATICS

1. For *A Century of Mathematics in America*, see above; the other sources are Karen Hunger Parshall and David E. Rowe, *The Emergence of the American Mathematical Community 1876–1900: J. J. Sylvester, Felix Klein, and E. H. Moore*, American and London Mathematical Societies, 1994, and Florian Cajori, *The Teaching and History of Mathematics in the United States*, Bureau of Information Circular 3, Washington Government Printing Office, 1890.
2. Harold L. Dorwart, "Mathematics at Yale in the nineteen twenties", *Century of Mathematics*, II, 87–97, on page 88.
3. Percey F. Smith, "The department of mathematics", *Century of Mathematics*, II, 121–26, on pages 121, 123.
4. William Aspray, "The emergence of Princeton as a world center for mathematical research, 1896–1939", *Century of Mathematics*, II, 195–215, on page 198.
5. *The Invisible University: Postdoctoral Education in the United States*, National Academy of Sciences Report (1969), on page 8.
6. Parshall and Rowe (note 1), on page 18.
7. Parshall and Rowe (note 1), on page 19.
8. Raymond Clare Archibald, *A Semicentennial History of the American Mathematical Society*, 1888–1938, Volume 1, American Mathematical Society (1938), 148.

CHAPTER 1: THE 1800s

In this chapter, we use the following abbreviation:

Sylvester Papers: J. J. Sylvester, *Collected Mathematical Papers of James Joseph Sylvester*, 4 volumes, Cambridge University Press, 1904–12.

1. For more information about Sylvester, see Karen Hunger Parshall's *James Joseph Sylvester: Life and Work in Letters* (referred to below as "*Letters*") and *James Joseph Sylvester: Jewish Mathematician in a Victorian World*, Johns Hopkins University Press, 1998 and 2006.
2. Letter from Benjamin Peirce to Daniel C. Gilman, 18 September 1875, Daniel C. Gilman Papers, Ms. 1, Special Collections Division, Milton S. Eisenhower Library, Johns Hopkins University, as quoted in Karen Hunger Parshall, "America's first school of mathematical research: James Joseph Sylvester at the Johns Hopkins University 1876–1883", *Archive for History of Exact Sciences* 38 (2) (1988), 153–96.
3. H. F. Baker, "Biographical notice", *Sylvester Papers*, Volume 4, xv–xxxvii, on page xxx.
4. Karen Hunger Parshall and David E. Rowe, "American mathematics comes of age: 1875–1900", *Century of Mathematics*, III, 3–28, on pages 8–9.
5. George P. Andrews, "The theory of partitions", *Encyclopaedia of Mathematics and Its Applications*, Volume 2, Addison-Wesley (1976), 14.
6. Letter from J. J. Sylvester to A. Cayley, 1 February 1883, J. J. Sylvester Papers, Library of St. John's College, Cambridge, Box 11.
7. E. Frankland, *Lecture Notes for Chemical Students*, London, 1866.
8. J. J. Sylvester, "Chemistry and algebra", *Nature* 17 (1877–78), 284 = *Sylvester Papers*, Volume 3, 103–4.
9. J. J. Sylvester, "On an application of the new atomic theory to the graphical representation of the invariants and covariants of binary quantics,—with three appendices", *American Journal of Mathematics* 1 (1878), 64–125, on pages 64, 87, and 109 = *Sylvester Papers*, Volume 3, 148–206, on pages 148, 169, and 190.
10. R. C. Archibald, "Unpublished Letters of James Joseph Sylvester and other new information concerning his life and work", *Osiris* 1 (1936), 85–154, on page 134.
11. *Graph Theory 1736–1936*, on page 67.
12. A. Cayley, "On the theory of the analytical forms called trees", and "On the theory of the analytical forms called trees—part II", *Philosophical Magazine* (*4*) 13 (1857), 172–76, and 18 (1859), 374–78 = *Cayley Papers*, Volume 3, 242–46, and Volume 4, 112–15.
13. A. Cayley, "On the mathematical theory of isomers", *Philosophical Magazine* (*4*) 47 (1874), 444–46 = *Cayley Papers*, Volume 9, 202–4.
14. A. Cayley, "On the analytical forms called trees, with application to the theory of chemical combinations", *Report of the British Association for the Advancement of Science* 45 (1875), 257–305, and "On the number of the univalent radicals C_nH_{2n+1}", *Philosophical Magazine* (*5*) 3 (1877), 34–35 = *Cayley Papers*, Volume 9, 427–60 and 544–45.
15. J. J. Sylvester, "On the mathematical question, what is a tree?", *Mathematical Questions with Their Solutions, from the "Educational Times"* 30 (1879), 52,

and [On the geometrical forms called trees], *Johns Hopkins University Circulars* 1 (1879–82), 202–3 = *Sylvester Papers*, Volume 3, 640–41.

16. [A. De Morgan, Unsigned review of W. Whewell, *The Philosophy of Discovery*], *The Athenaeum*, No. 1694 (April 14, 1860), 501–3.
17. K. Appel and W. Haken, "Every planar map is four colorable, Part I: Discharging", and K. Appel, W. Haken, and J. Koch, "Every planar map is four colorable, Part II: Reducibility", *Illinois Journal of Mathematics* 21 (1977), 429–90, and 491–567. For more information on the four color problem, see *Four Colors Suffice*.
18. [Questions asked by Prof. Cayley F.R.S.], *Proceedings of the London Mathematical Society* 9 (1877–78), 148, and [Report of meeting on 11 July 1878], *Nature* (1878), 294. The earliest known reference to the problem is a paragraph that appeared in *The Athenaeum* of June 10, 1854; see Brandan D. McKay, "A note on the history of the four-colour conjecture", *Journal of Graph Theory* 72 (2013), 361–63, and *Four Colors Suffice*, on page 15.
19. A. Cayley, "On the colouring of maps", *Proceedings of the Royal Geographical Society* (New Series) 1 (1879), 259–61 = *Cayley Papers*, Volume 11, 7–8.
20. A. B. Kempe, [Notes], *Nature* 20 (17 July 1879), 275.
21. A. B. Kempe, "On the geographical problem of the four colours", *American Journal of Mathematics* 2 (1879), 193–200.
22. A. B. Kempe, [Untitled], *Proceedings of the London Mathematical Society* 10 (1878–79), 229–31, and "How to colour a map with four colours", *Nature* 21 (1879–80), 399–400.
23. Roger Cooke and V. Frederick Rickey, "W. E. Story of Hopkins and Clark", *Century of Mathematics*, III, 29–76, on page 35.
24. See Archibald (note 10), on page 139.
25. William E. Story, "Note on the preceding paper [by Kempe]", *American Journal of Mathematics* 2 (1879), 201–4.
26. Letter from J. J. Sylvester to Daniel C. Gilman, 22 July 1880, Daniel C. Gilman Papers, Ms. 1 (see note 2). It appears in full in *Letters* (note 1), on pages 194–98.
27. Max H. Fisch and Jackson I. Cope, "Peirce at the Johns Hopkins University", *Studies in the Philosophy of Charles Sanders Peirce* (ed. P. P. Wiener and F. H. Young), Cambridge, 1952.
28. Letter from J. J. Sylvester to Felix Klein, 17 January 1884, cited by Karen H. Parshall and David E. Rowe in *J. J. Sylvester, Felix Klein and E. H. Moore, The Emergence of the American Mathematical Research Community 1876–1900*, American Mathematical Society, 1994. See also *Letters* (note 1), on page 235.
29. Francesco Cordasco, *Daniel Coit Gilman and the Protean Ph.D. The Shaping of American Graduate Education*, E. J. Brill, 1960.
30. See Cooke and Rickey (note 23), on page 49.
31. R. M. Martin (ed.), *Studies in the Scientific and Mathematical Philosophy of Charles S Peirce: Essays by Carolyn Eisele*, Mouton, 1971; Peirce's interest in the four color problem is on pages 216–22. See also Norman L. Biggs, E. Keith Lloyd, and Robin J. Wilson, "C. S. Peirce and De Morgan and the four-colour conjecture", *Historia Mathematica* 4 (1976), 215–16.
32. See Martin (note 31).
33. See Martin (note 31).

34. [Report of meeting of Scientific Association, 3 December 1879], *Johns Hopkins University Circulars* 1 (2) (January 1880), 16.
35. "The beginnings of the American Mathematical Society: Reminiscences of Thomas Scott Fiske", *Century of Mathematics*, I, 13–17, on pages 15–16.
36. [Report of meeting of Scientific Session, 15 November 1899], *Report of the National Academy of Sciences* (1899), 12–13.
37. The letter from W. Story to C. S. Peirce appears in Martin (note 31), on page 359; see also *Four Colors Suffice*, on page 70.
38. Bertrand Russell, *Wisdom of the West*, Crescent (1959), 276.

INTERLUDE A: GRAPH THEORY IN EUROPE 1

In this chapter, we use the following abbreviation:

Tait Papers: P. G. Tait, *Scientific Papers*, 2 volumes, Cambridge University Press, 1898–1900.

1. P. G. Tait, "On the colouring of maps" and "Remarks on the previous communication [by Frederick Guthrie]", *Proceedings of the Royal Society of Edinburgh* 10 (1879–80), 501–3 and 729, and "Note on a theorem in the geometry of position", *Transactions of the Royal Society of Edinburgh* 29 (1878–80), 657–60 = *Tait Papers*, Volume 1, 408–11.
2. P. G. Tait, "Listing's *Topologie*", *Philosophical Magazine* (*5*) 17 (1884), 30–46 = *Tait Papers*, Volume 2, 85–98.
3. T. P. Kirkman, "Question 6610 with solution by the proposer", *Mathematical Questions and Solutions from the "Educational Times"* 35 (1881), 112–16.
4. P. J. Heawood, "Map-colour theorem", *Quarterly Journal of Pure and Applied Mathematics* 24 (1890), 332–38.
5. See *Graph Theory 1736–1936*, on pages 107–8, and *Four Colors Suffice*, on page 102.
6. P. J. Heawood, "On the four-colour map theorem", *Quarterly Journal of Pure and Applied Mathematics* 29 (1898), 270–85.
7. S.-A.-J. L'Huilier, "Mémoire sur la polyédrométrie, contenant une demonstration directe du théorème d'Euler sur les polyèdres, et un examen des diverses exceptions auxquelles ce théorème est assujetti", *Annales de Mathématiques Pures et Appliquées* 3 (1812–13), 169–89.
8. Later papers on map coloring by P. J. Heawood include "On extended congruences connected with the four-colour map theorem", "Failures in congruences connected with the four-colour map theorem", and "Map-colour theorem", *Proceedings of the London Mathematical Society* (*2*) 33 (1932), 253–86; 40 (1936), 189–202; and 51 (1949), 161–75.
9. Julius Petersen, "Die Theorie der regulären Graphs", *Acta Mathematica* 15 (1891), 193–220.
10. See G. Sabidussi, "Correspondence between Sylvester, Petersen, Hilbert and Klein on invariants and the factorisation of graphs 1889–1891", *Discrete Mathematics* 100 (1992), 99–155.
11. See H. R. Brahana, "A proof of Petersen's theorem", *Annals of Mathematics* (*2*) 19 (1917–18), 59–63; Alfred Errera, "Une démonstration du théorème

de Petersen", *Mathesis* 36 (1922) 56–61; and O. Frink, "A proof of Petersen's theorem", *Annals of Mathematics* (2) 27 (1925–26), 491–93.
12. Julius Petersen, [Sur le théorème de Tait], *L'Intermédiaire des Mathématiciens* 5 (1898), 225–27.
13. A. B. Kempe, "A memoir on the theory of mathematical form", *Philosophical Transactions of the Royal Society of London* 177 (1886), 1–70, on page 11.
14. Julius Petersen, [Réponse à question 360], *L'Intermédiaire des Mathématiciens* 6 (1899), 36–38.
15. L. Heffter, "Über das Problem der Nachbargebiete", *Mathematiche Annalen* 38 (1891), 477–508.
16. The problem was named the "thread problem" by David Hilbert and Stefan Cohn-Vossen in their book *Anschauliche Geometrie*, Springer, 1932, which was based on a series of lectures given in 1920–21 by Hilbert; the book was later published in English as *Geometry and the Imagination*.
17. H. Tietze, *Famous Problems of Mathematics. Solved and Unsolved Mathematical Problems from Antiquity to Modern Times*, Graylock Press (Baltimore), 1965.
18. H. Tietze, "Einige Bemerkungen über das Problem des Kartenfärbens auf einseitigen Flächen", *Jahresbericht der Deutschen Mathematiker-Vereinigung* 19 (1910), 155–59.
19. Constance Reid, *Hilbert*, Springer (1970), 92–93.

CHAPTER 2: THE 1900s AND 1910s

In this chapter, we use the following abbreviation:

Birkhoff Papers: G. D. Birkhoff, *Collected Mathematical Papers*, 3 volumes, American Mathematical Society, 1950.

1. P. Wernicke, [On the solution of the map-color problem], *Bulletin of the American Mathematical Society* 4 (1897–98), 5.
2. P. Wernicke, "Über den kartographischen Vierfarbensatz", *Mathematiche Annalen* 58 (1904), 413–26.
3. S. Mac Lane, "Mathematics at the University of Chicago: A brief history", *Century of Mathematics*, II, 127–54, on p. 129.
4. Steve Batterson, "The vision, insight, and influence of Oswald Veblen", *Notices of the American Mathematical Society* 54 (2007), 606–18.
5. Oswald Veblen, "An application of modular equations in analysis situs", *Annals of Mathematics* (2) 14 (1912–13), 86–94.
6. For Henri Poincaré's use of matrices, see his "Second complément à l'Analysis Situs", *Proceedings of the London Mathematical Society* 32 (1900), 277–308 = *Oeuvres de Henri Poincaré*, Volume 6, Gauthier-Villars (Paris), 338–72.
7. Oswald Veblen, *The Cambridge Colloquium 1916: Part II, Analysis Situs*, American Mathematical Society Colloquium Lectures, Volume V, New York, 1922.
8. G. R. Kirchhoff, "Über die Auflösung der Gleichungen, auf welche man bei der Untersuchung der linearen Vertheilung galvanischer Ströme geführt wird", *Annalen der Physik und Chemie* 72 (1847), 497–508 = *Gesammelte Anhandlungen von G. Kirchhoff*, Barth (Leipzig) (1882), 22–33.

9. Marston Morse, "George David Birkhoff and his mathematical work", *Bulletin of the American Mathematical Society* 52 (1946), 357–91.
10. For an outline of life at Harvard University at this time, see Garrett Birkhoff, "Mathematics at Harvard, 1836–1944", *Century of Mathematics*, II, 3–58.
11. Interview with Garrett Birkhoff in *Mathematical People: Profiles and Interviews* (ed. D. J. Albers and G. L. Alexanderson), Birkhäuser (Boston) (1985), 12–13.
12. Hassler Whitney, "Moscow 1935: Topology moving toward America", *Century of Mathematics*, I, 97–117, on page 99.
13. George D. Birkhoff, "A determinant formula for the number of ways of coloring a map", *Annals of Mathematics* (2) 14 (1912–13), 42–46 = *Birkhoff Papers*, Volume 3, 1–5.
14. G. D. Birkhoff, "On the number of ways of coloring a map", *Proceedings of the Edinburgh Mathematical Society* (2) 2 (1930), 83–91 = *Birkhoff Papers*, Volume 3, 20–28.
15. G. D. Birkhoff, "On the polynomial expressions for the number of ways of coloring a map", *Annali della Scuola Normale Superiore di Pisa* (2) 3 (1934), 85–104 = *Birkhoff Papers*, Volume 3, 29–47.
16. George D. Birkhoff, "The reducibility of maps", *American Journal of Mathematics* 35 (1913), 115–28 = *Birkhoff Papers*, Volume 3, 6–19.
17. D. A. Rothrock (ed.), "American mathematicians in war service", *Century of Mathematics*, I, 269–73.

CHAPTER 3: THE 1920s

1. D. V. Widder, "Some mathematical reminiscences", *Century of Mathematics*, I, 79–83.
2. Dirk J. Struik, "The MIT Department of Mathematics during its first seventy-five years: Some recollections", *Century of Mathematics*, III, 163–77, on page 171.
3. Philip Franklin, "The four color problem", *American Journal of Mathematics* 44 (1922), 225–36.
4. Philip Franklin, "The electric currents in a network", *Journal of Mathematics and Physics* 4 (1925), 97–102.
5. Ronald M. Foster, "Geometrical circuits of electrical networks", *Transactions of the American Institute of Electrical Engineers* 51 (1932), 309–17.
6. I. N. Kagno, "A note on the Heawood color formula", *Journal of Mathematical Physics* 14 (1935), 228–31; H.S.M. Coxeter, "The map-coloring of unorientable surfaces", *Duke Mathematical Journal* 10 (1943), 293–304; and R. C. Bose, "On the construction of balanced incomplete block designs", *Annals of Eugenics* 9 (1939), 353–99.
7. Philip Franklin, "A six color problem", *Journal of Mathematics and Physics* 13 (1934), 363–69.
8. H. R. Brahana, "A proof of Petersen's theorem", *Annals of Mathematics* (2) 19 (1917–18), 59–63.

9. H. R. Brahana, "The four color problem", *American Mathematical Monthly* 30 (1923), 234–43.
10. H. R. Brahana, "Systems of circuits of two-dimensional manifolds", *Annals of Mathematics* 30 (1923), 234–43.
11. H. R. Brahana, "Regular maps on an anchor ring", *American Journal of Mathematics* 48 (1926), 225–40.
12. G. A. Miller, "On the groups generated by two operators of order two and three respectively whose product is of order six", *Quarterly Journal of Pure and Applied Mathematics* 33 (1901), 76–79.
13. R. P. Baker, "Cayley diagrams on the anchor ring", *American Journal of Mathematics* 53 (1931), 645–69.
14. For information about Redfield's life and work, see E. Keith Lloyd's articles, "J. Howard Redfield 1879–1944", *Journal of Graph Theory* 8 (1984), 195–203, and "Redfield's papers and their relevance to counting isomers and isomerizations", *Discrete Applied Mathematics* 19 (1988), 289–304, and his chapter on "Enumeration (18th–20th centuries)" in *Combinatorics: Ancient & Modern* (ed. R. Wilson and J. J. Watkins), Oxford University Press (2013), 284–307.
15. Percy A. MacMahon, *Combinatory Analysis*, 2 volumes, Cambridge University Press, 1915–16; reprinted in one volume by Chelsea Publishing in 1960.
16. J. Howard Redfield, "The theory of group-reduced distributions", *American Journal of Mathematics* 49 (1927), 433–55.
17. G. Pólya, *How to Solve It*, Princeton University Press, 1945.
18. G. Pólya, "Kombinatorische Anzahlbestimmungen für Gruppen, Graphen und chemische Verbindungen", *Acta Mathematica* 68 (1937), 308–416.
19. G. Pólya and R. C. Read, *Combinatorial Enumeration of Groups, Graphs, and Chemical Compounds*, Springer-Verlag, 1987.
20. C. N. Reynolds Jr., "On the problem of coloring maps in four colors, I, II", *Annals of Mathematics* (*2*) 28 (1926–27), 1–15 and 477–92.
21. Clarence N. Reynolds Jr., "Circuits upon polyhedra", *Annals of Mathematics* (*2*) 33 (1932), 367–72.
22. Philip Franklin, "Note on the four color problem", *Journal of Mathematics and Physics* 16 (1938), 172–84.
23. I. Ratib and C. E. Winn, "Généralisation d'une réduction d'Errera dans le problème des quatre couleurs", *Comptes Rendus Congress of International Mathematicians, Oslo*, Volume 2 (1936), 131–33.
24. C. E. Winn, "A class of coloration in the four color problem", *American Journal of Mathematics* 59 (1937), 515–28; "Sur quelques réductibilités dans la théorie des cartes", *Comptes Rendus de l'Académie des Sciences* 205 (1937), 352–54; and "On certain reductions in the four color problem", *Journal of Mathematics and Physics* 16 (1938), 159–71.
25. C. E. Winn, "Sur l'historique du problème des quatre couleurs", *Bulletin of the Institute of Egypt* 20 (1939), 191–92.
26. C. E. Winn, "On the minimum number of polygons in an irreducible map", *American Journal of Mathematics* 62 (1940), 406–16.
27. Philip Franklin, "The four color problem", *Galois Lectures, Scripta Mathematica Library* 6 (1939), 149–56 and 197–210.

INTERLUDE B: GRAPH THEORY IN EUROPE 2

1. Dénes König, *Mathematical Recreations I, II*, Budapest, 1902, 1905.
2. Dénes König, "A térképszinezésről" (On map coloring), *Matematikai és Fizikai Lapok* 14 (1905), 193–200.
3. Tibor Gallai, "The life and scientific work of Dénes König (1884–1944)", *Linear Algebra and its Applications* 21 (1978), 189–205.
4. Dénes König, "Line systems on two-sided surfaces" and "The genus of line systems" (in Hungarian), *Mathematikai és Természettudományi Értesitő* 29 (1911), 112–17 and 345–50.
5. Dénes König, "Grafók és alkalmazásuk a determinánsok és halmazok elméletében", *Mathematikai és Természettudományi Értesitő* 34 (1916), 104–19= "Über Graphen und ihre Anwendungen auf Determinantentheorie und Mengenlehre" (On graphs and their applications in determinant theory and set theory), *Mathematische Annalen* 77 (1916), 453–65.
6. Dénes König, *The Elements of Analysis Situs* (in Hungarian), Budapest, 1918.
7. Dénes König and S. Valkó, "Many-valued maps of sets", *Mathematikai és Természettudományi Értesitő* 42 (1926), 173–76, and D. König, "Über eine Schlussweise aus dem Endlichen ins Unendliche", *Acta Litterarum Scientiarum Szeged* 3 (1927), 121–30.
8. Dénes König, "Graphok és matrixok" (Graphs and matrices), *Matematikai és Fizikai Lapok* 38 (1931), 116–19.
9. E. Egerváry, "Matrixok kombinatorius tulajdonságairól" (Combinatorial properties of matrices), *Matematikai és Fizikai Lapok* 38 (1931), 16–28.
10. Dénes König, "Über trennende Knotenpunke in Graphen (nebst Anwendungen auf Determinanten und Martizen)", *Acta Litterarum Scientiarum Szeged* 6 (1933), 155–79.
11. Dénes König, *Theorie der endlichen und unendlichen Graphen*, Akademischen Verlagsgesellschaft, (Leipzig), 1936; reprinted by Chelsea (New York), 1950.
12. Dénes König, *Theory of Finite and Infinite Graphs* (translated by R. McCoart, with a commentary by W. T. Tutte), Birkhäuser, 1990.
13. Alfred Errera, *Du Coloriage des Cartes et de Quelques Questions d'Analysis Situs*, Falk Fils (Brussels) and Gauthier-Villars (Paris), 1921.
14. Alfred Errera, "Une démonstration du théorème de Petersen", *Mathesis* 36 (1922), 56–61.
15. Henry E. Dudeney, "Perplexities", *Strand Magazine* 46, No. 271 (July 1913), 110, and No. 272 (August 1913), 221.
16. Alfred Errera, "Un théorème sur les liaisons", *Comptes Rendus* 177 (1923), 489–91.
17. Alfred Errera, "Quelques remarques sur le problème des quatre couleurs", *Proceedings of the International Congress of Mathematicians, Toronto*, Volume 1 (1924), 693–94.
18. Alfred Errera, "Exposé historique du problème des quatre couleurs", *Periodico di Matematiche* (*4*) 7 (1927), 20–41.
19. A. Errera, "Une contribution au problème des quatre couleurs", *Bulletin de la Société Mathématiques de France* 53 (1925), 42–55.
20. Later papers of Alfred Errera, which indicate the range of his interests, include "Sur le problème des quatre couleurs, I, II", *Académie Royale Belgique*

Bulletin de la Classe-des Sciences (*5*) 33 (1947), 807–21, and 34 (1948), 65–84; "Sur un théorème de M. Whitney, un problème de Lebesgue et les réseaux de Tait", *IIIe Congrès National des Sciences*, 2 (1950), 51–55; "Sur la classification des polyèdres de genre zero", *Bulletin de la Société Mathématiques de Belgique* (1952), 51–66; "Une vue d'ensemble sur le problème des quatre couleurs", *University e Politecnico, Torino. Rendiconti Seminario Matematico* 11 (1952), 5–19; and "Sur les polyèdres de genre zero", *Università di Roma, 1st Nazionale alta Matematica, Rendiconti de Matematica e delle Sue Applicazioni* 11 (1952), 315–22.

21. André Sainte-Laguë, *Les Réseaux* (*ou graphes*), Mémorial des Sciences Mathématiques 18, Gauthier-Villars (Paris), 1926.
22. André Sainte-Laguë, *The Zeroth Book of Graph Theory: An Annotated Translation of Les Réseaux* (*ou graphes*); English translation by M. C. Golumbic, Springer-Verlag, 2021.
23. André Sainte-Laguë, *Géométrie de Situation et Jeux*, Mémorial des Sciences Mathématiques 41, Gauthier-Villars (Paris), 1929.
24. Karl Menger, "Zur allgemeinen Kurventheorie", *Fundamenta Mathematicae* 10 (1927), 96–115.
25. Karl Menger, *Kurventheorie*, Teubner, 1932.
26. P. Hall, "On representatives of subsets", *Journal of the London Mathematical Society* 10 (1935), 26–30; and Paul R. Halmos and Herbert E. Vaughan, "The marriage problem", *American Journal of Mathematics* 72 (1950), 214–15.
27. Casimir Kuratowski, "Sur le problème des courbes gauches en topologie", *Fundamenta Mathematicae* 15 (1930), 271–83.
28. K. Kuratowski, "My personal recollections connected with the research on some topological problems", *Colloquio Internazionale sulle Theorie Combinatorie* (*Rome 1973*), Volume I, Atti die Convegni Lincei 17, Accademia Nazionale die Lincei (1976), 43–47.
29. K. Menger, "Über plättbare Dreiergraphen und Potenzen nichtplättbarer Graphen", *Anzeiger der Akademie der Wissenschaften in Wien* 67 (1930), 85–86 = *Ergebnisse eines Mathematischen Kolloquiums* 2 (1930), 30–31.
30. Orrin Frink and Paul A. Smith, "Abstract 179", *Bulletin of the American Mathematical Society* 36 (1930), 214.
31. Orrin Frink, letter to Robin Wilson, 1974; see *Graph Theory 1736–1936*, on page 148.
32. Orrin Frink, letter to J. W. Kennedy, L. V. Quintas, and M. M. Sysło, 1981 (see note 33).
33. John W. Kennedy, Louis V. Quintas, and Maciej M. Sysło, "The theorem on planar graphs", *Historia Mathematica* 12 (1985), 356–68.

CHAPTER 4: THE 1930s

1. W. T. Tutte, "Commentary", in Dénes König, *Theory of Finite and Infinite Graphs* (translated by R. McCoart), Birkhäuser (1990), 1; the phrase "the slums of Topology" has been attributed to the topologist J. H. C. Whitehead.
2. A biography of Hassler Whitney that presents a revealing portrait of a fascinating personality and a giant of 20th-century mathematics is Keith Kendig's *Never a Dull Moment: Hassler Whitney, Mathematics Pioneer*, MAA Press, 2018.

3. Hassler Whitney, "Moscow 1935: Topology moving toward America", *Century of Mathematics*, I, 97–117, on pages 106–7.
4. Hassler Whitney, "The coloring of graphs", *Proceedings of the National Academy of Sciences* 17 (1931), 122–25, and *Annals of Mathematics* (2) 33 (1932), 688–718.
5. Personal communication by Hassler Whitney at a meeting with Robin Wilson at the Institute for Advanced Study, Princeton, on January 29, 1975.
6. Hassler Whitney, "A logical expansion in mathematics", *Bulletin of the American Mathematical Society* 38 (1932), 572–79.
7. Hassler Whitney, "A theorem on graphs", *Annals of Mathematics* (2) 32 (1931), 378–90.
8. See note 5.
9. Hassler Whitney, "A numerical equivalence of the four color problem", *Monatshefte für Mathematik und Physik* 45 (1936), 207–13.
10. Hassler Whitney, "Non-separable and planar graphs", *Proceedings of the National Academy of Sciences* 17 (1931), 125–27, and *Transactions of the American Mathematical Society* 34 (1932), 339–62.
11. Hassler Whitney, "Planar graphs", *Fundamenta Mathematicae* 21 (1933), 73–84.
12. Hassler Whitney, "On the abstract properties of linear dependence", *American Journal of Mathematics* 57 (1935), 509–33.
13. Saunders Mac Lane, "Some interpretations of abstract linear dependence in terms of projective geometry", *American Journal of Mathematics* 58 (1936), 236–40.
14. B. L. van der Waerden, *Moderne Algebra*, second edition, Springer, 1937.
15. R. Rado, "A theorem on independence relations", *Quarterly Journal of Mathematics (Oxford)* 13 (1942), 83–89.
16. See Tutte (note 1), on page 1.
17. Hassler Whitney, "Congruent graphs and the connectivity of graphs", *American Journal of Mathematics* 54 (1932), 150–68.
18. Hassler Whitney, "2-isomorphic graphs", *American Journal of Mathematics* 55 (1933), 245–54.
19. Hassler Whitney, "On the classification of graphs", *American Journal of Mathematics* 55 (1933), 236–44.
20. Ronald M. Foster, "Geometrical circuits of electrical networks", *Transactions of the American Institute of Electrical Engineers* 51 (1932), 309–17.
21. Hassler Whitney, "A set of topological invariants for graphs", *American Journal of Mathematics* 55 (1933), 231–35.
22. Whitney's papers on these subjects are: "Analytic extensions of differentiable functions defined in closed sets", *Transactions of the American Mathematical Society* 36 (1934), 63–89; "Differentiable manifolds", *Annals of Mathematics* (2) 37 (1936), 645–80; "The self-intersections of a smooth n-manifold in $2n$-space", *Annals of Mathematics* (2) 45 (1944), 220–46; "On singularities of mappings of Euclidean spaces. I. Mappings of the plane into the plane", *Annals of Mathematics* (2) 62 (1955), 374–410; and "Tangents to an analytic variety", *Annals of Mathematics* (2) 81 (1965), 496–549. Further information about these topics can be found in Chapters 11–14 and 16–17 of Kendig's book (note 2).

23. Saunders Mac Lane, "Some unique separation theorems for graphs", *American Journal of Mathematics* 57 (1935), 805–20.
24. Saunders Mac Lane, "A combinatorial condition for planar graphs", *Fundamenta Mathematicae* 28 (1937), 22–32.
25. Saunders Mac Lane, "A structural characterization of planar combinatorial graphs", *Duke Mathematical Journal* 3 (1937), 460–72.
26. See Ivan Niven, "The threadbare thirties", *Century of Mathematics*, I, 209–29.
27. From 1923 to 1928 the International Board of the Rockefeller Foundation initiated a scheme of one-year fellowships that enabled distinguished scholars in the natural sciences to visit institutions in other countries. George Birkhoff was very involved with this scheme, and in 1926 paid a lengthy visit to Europe to recruit outstanding young scholars in difficult circumstances to positions in the United States. Later, in the 1930s, he had a very different attitude. For further information, see Reinhard Siegmund-Schultze, *Rockefeller and the Internationalization of Mathematics between the Two World Wars*, Springer Basel AG, 2001, and Karen Hunger Parshall, *The New Era in American Mathematics, 1920–1950*, Princeton University Press, 2022.
28. E. R. Murrow, Memo on "Displaced American scholars", *Archives of the Rockefeller Foundation*, Rockefeller Archive Centre, New York, in archive 2/717/92/731, 1933.
29. Letter from O. Veblen to R.G.D. Richardson, 6 May 1935, American Mathematical Society. See also Nathan Reingold, "Refugee mathematicians in the United States of America, 1933–1941: Reception and reaction", *Annals of Science* 38 (1981), 313–38, reprinted in *Century of Mathematics*, I, pp. 175–200, on page 178.
30. See Reingold (note 29), on page 184.
31. Letter from Harlow Shapley to Hermann Weyl, 24 May 1939, Oswald Veblen Papers, Box 29, Manuscript Division, Library of Congress; see also Reingold (note 29), on page 196.
32. Letter from G. D. Birkhoff to R.G.D. Richardson, 18 May 1934, R.G.D. Richardson Papers, Archives, Brown University; see also Reingold (note 29), on page 183.
33. See Reingold (note 29), on page 184.
34. Lipman Bers, "The migration of European mathematicians to America", *Century of Mathematics*, I, pp. 231–43, on pages 235–36.
35. Saunders Mac Lane, "Jobs in the 1930s and the views of George D. Birkhoff", *The Mathematical Intelligencer* 16 (Summer 1994), 9–10.

CHAPTER 5: THE 1940s AND 1950s

1. G. D. Birkhoff and D. C. Lewis, "Chromatic polynomials", *Transactions of the American Mathematical Society* 60 (1946), 355–451.
2. These letters between Daniel Lewis, George Birkhoff, and Clarence Reynolds in August and September 1942 are in the Harvard University archives, Ref. 4213.4.5, Correspondence c.1937–43. The authors of this volume are grateful to June Barrow-Green for obtaining copies for our use.
3. See Birkhoff and Lewis (note 1), on p. 358.

4. W. T. Tutte, *Graph Theory as I Have Known It*, Clarendon Press (Oxford) (1998/2012), 135.
5. Arthur Bernhart, "Six-rings in minimal five-color maps", *American Journal of Mathematics* 69 (1947), 391–412.
6. Arthur Bernhart, "Another reducible edge configuration", *American Journal of Mathematics* 70 (1948), 144–46.
7. Frank R. Bernhart, "A three and five color theorem", *Proceedings of the American Mathematical Society* 52 (1975), 493–98.
8. Frank R. Bernhart, "The four color theorem proved by multi-linear algebra???", *Proceedings of the Conference on Algebraic Aspects of Combinatorics*, University of Toronto (1975), 219–25.
9. Henri Lebesgue, "Quelques conséquences simples de la formule d'Euler", *Journal de Mathématiques Pures et Appliquées* 9 (1940), 27–43.
10. I. N. Kagno, "Perfect subdivisions of surfaces", *Journal of Mathematics and Physics* 17 (1938), 76–111.
11. Papers of I. N. Kagno from the 1930s include "A note on the Heawood color formula", "The triangulation of surfaces and the Heawood color formula", and "The mapping of graphs on surfaces", *Journal of Mathematics and Physics* 14 (1935), 228–31, 15 (1936), 179–86, and 16 (1937), 46–75.
12. R. Frucht, "Herstellung von Graphen mit vorgegebener abstrakten Gruppe", *Compositio Mathematica* 6 (1938), 239–50, and "Graphs for degree three with a given abstract group", *Canadian Journal of Mathematics* 1 (1949), 365–78.
13. A. Cayley, "The theory of groups: graphical representation", *American Journal of Mathematics* 1 (1878), 174–76 = *Cayley Papers*, Volume 10, 403–5.
14. I. N. Kagno, "Linear graphs of degree ≤ 6 and their groups", *American Journal of Mathematics* 68 (1946), 505–20, and 69 (1947), 872.
15. I. N. Kagno, "Desargues' and Pappus' graphs and their groups", *American Journal of Mathematics* 69 (1947), 859–63.
16. Sister Petronia Van Straten, Abstract 345: "Toroidal and non-toroidal graphs", *Bulletin of the American Mathematical Society* 52 (1846), 831; the full paper is "The topology of the configuration of Desargues and Pappus", *Reports of a Mathematical Colloquium* 8, University of Notre Dame (1948), 1–17.
17. A. Cayley, "On the theory of the analytical forms called trees", *Philosophical Magazine* (4) 13 (1857), 172–76 = *Cayley Papers*, Volume 3, 242–46.
18. A. Cayley, "On the analytical forms called trees", *American Journal of Mathematics* 4 (1881), 266–68 = *Cayley Papers*, Volume 11, 365–67.
19. Richard Otter, "The number of trees", *Annals of Mathematics* 49 (1948), 583–99.
20. Claude E. Shannon, "A theorem on coloring the lines of a network", *Journal of Mathematics and Physics* 28 (1949), 148–51.
21. V. G. Vizing, "The chromatic class of a multigraph", *Kibernetika* (Kiev) 3 (1965), 29–39, and *Cybernetics* 1 (1965), 32–41; and "Critical graphs with a given chromatic class" (in Russian), *Diskretnyi Analiz* 5 (1965), 9–17.
22. See Tutte (note 4).
23. D. H. Younger, "William Thomas Tutte. 14 May 1917–2 May 2002", *Biographical Memoirs of Fellows of the Royal Society* (2012), 282–97, and Arthur M.

Hobbs and James G. Oxley, "William T. Tutte, 1917–2001", *Notices of the American Mathematical Society* 51 (2004), 320–30.

24. W. T. Tutte, *Selected Papers of W. T. Tutte*, Vols. I, II (ed. D. McCarthy and R. G. Stanton), Charles Babbage Research Centre (Winnipeg), 1979.
25. See Younger (note 23), on p. 287; the reference is to W. W. Rouse Ball, *Mathematical Recreations and Essays* (originally entitled *Mathematical Recreations and Problems of Past and Present Times*), Macmillan, 1892.
26. See Tutte (note 4), on page 1.
27. See Younger (note 23), on page 288.
28. G.B.B.M. Sutherland and W. T. Tutte, "Absorption of polymolecular films in the infra-red", *Nature* 144 (1939), 707.
29. See Younger (note 23), on page 292.
30. Graham Farr and James Oxley, "Contributions of W. T. Tutte to matroid theory", *2017 Matrix Annals* (ed. J. de Gier, C. E. Praeger, and T. Tao), Springer (2019), 343–61.
31. W. T. Tutte, "Graph-polynomials", *Advances in Applied Mathematics* 32 (2004), 5–9, on page 5.
32. R. L. Brooks, C.A.B. Smith, A. H. Stone, and W. T. Tutte, "The dissection of rectangles into squares", *Duke Mathematical Journal* 7 (*1*) (1940), 312–40.
33. Z. Moroń, "O rozkladąch prostokatów na kwadraty" (On the dissection of a rectangle into squares), *Przegląd Matematyczno-Fizycny* 3 (1925), 152–53.
34. R. Sprague, "Beispiel einer Zerlegung des Quadrats in lauter verschiedene Quadrate", *Mathematische Zeitschrift* 45 (1939), 607–8.
35. A.J.W. Duijvestijn, "Simple perfect squared square of lowest order", *Journal of Combinatorial Theory* (*B*) 25 (2) (1978), 240–43.
36. R. L. Brooks, "On colouring the nodes of a network", *Proceedings of the Cambridge Philosophical Society* 37 (1941), 194–97.
37. W. T. Tutte, "Squaring the square", *Canadian Journal of Mathematics* 2 (1950), 197–209.
38. W. T. Tutte, "On Hamiltonian circuits", *Journal of the London Mathematical Society* 21 (1946), 98–101.
39. W. T. Tutte, "A theorem on planar graphs", *Transactions of the American Mathematical Society* 82 (1956), 99–116.
40. W. T. Tutte, "The factorization of linear graphs", *Journal of the London Mathematical Society* 22 (1947), 107–11.
41. T. Gallai, "On factorisation of graphs", *Acta Mathematica Academiae Scientiarum Hungaricae* 1 (1950), 133–53, and F. G. Maunsell, "A note on Tutte's paper "The factorization of linear graphs", *Journal of the London Mathematical Society* 27 (1952), 127–28. See also T. Gallai, "Neuer Beweis eines Tutte'schen Satzes" (A new proof of Tutte's theorem), *A Magyar Tudományos Akadémia Matematikai Kutató Intézetének Közlemenyei* 9 (1963), 135–39.
42. W. T. Tutte, "The factors of graphs", *Canadian Journal of Mathematics* 4 (1952), 314–28.
43. W. T. Tutte, "A short proof of the factor theorem for finite graphs", *Canadian Journal of Mathematics* 6 (1954), 347–52.
44. W. T. Tutte, "A homotopy theorem for matroids, I, II" and "Matroids and graphs", *Transactions of the American Mathematical Society* 88 (1958), 144–60, 161–74, and 90 (1959), 527–52.

45. G. Fano, "Sui postulari fondamentali della geometria proiectiva", *Giornale di Matematiche* 30 (1892), 106–32.
46. K. Wagner, "Über eine Erweiterung eines Satzes von Kuratowski", *Deutsche Mathematik* 2 (1937), 280–85, and F. Harary and W. T. Tutte, "A dual form of Kuratowski's theorem", *Canadian Mathematical Bulletin* 8 (1965), 17–20.
47. W. T. Tutte, "A ring in graph theory", *Proceedings of the Cambridge Philosophical Society* 43 (1947), 26–40.
48. See Tutte (note 31), on page 7.
49. W. T. Tutte and G. Berman, "The golden root of a chromatic polynomial", *Journal of Combinatorial Theory* 6 (1969), 301–2.
50. W. T. Tutte, "On chromatic polynomials and the golden ratio", *Journal of Combinatorial Theory* 9 (1970), 289–96.
51. W. T. Tutte, "More about chromatic polynomials and the golden ratio", *Combinatorial Structures and Their Applications* (ed. R. K. Guy et al.), Gordon and Breach (1970), 439–53.
52. Ralph G. Stanton, Foreword to *Selected Papers of W. T. Tutte* (note 24).
53. M. Fleury, "Deux problèmes de géométrie de situation", *Journal de Mathématiques Élementaires* (*2*) 2 (1883), 257–61; G. Tarry, "Le problème des labyrinthes", *Nouvelles Annales de Mathématiques* (*3*) 14 (1895), 187–90; and É. Lucas, *Récréations Mathématiques*, Volume 1, Gauthier-Villars (Paris), 1882.
54. Alexander Schrijver, "On the history of combinatorial optimization (till 1960)", *Discrete Optimization* (ed. K. Aardal, G. L. Nemhauser, and R. Weismantel), *Handbooks in Operations Research and Management Science* 12, North-Holland (2005), 1–68; and *Combinatorial Optimization: Polyhedra and Efficiency*, Springer, 2003.
55. Dénes König, "Graphok és matrixok" (Graphs and matrices), *Matematikai és Fizikai Lapok* 38 (1931), 116–19.
56. E. Egerváry, "Matrixok kombinatorius tulajdonságairól" (Combinatorial properties of matrices), *Matematikai és Fizikai Lapok* 38 (1931), 16–28.
57. H. W. Kuhn, "The Hungarian Method for the assignment problem", *Naval Research Logistics* 2 (1955), 83–97.
58. András Frank, "On Kuhn's Hungarian Method—A tribute from Hungary", *Naval Research Logistics* 52 (2005), 2–5.
59. H. W. Kuhn, "On the origin of the Hungarian method", *History of Mathematical Programming—A Collection of Personal Reminiscences* (ed. J. K. Lenstra et al.), Centrum Wiskunde & Informatica (CWI) (Amsterdam) (1991), 77–81.
60. Frank L. Hitchcock, "The distribution of a product from several sources to numerous localities", *Journal of Mathematics and Physics* 20 (1941), 224–30.
61. G. B. Dantzig, "Maximization of a linear function of variables subject to linear inequalities", and "Application of the simplex method to a transportation problem", *Activity Analysis of Production and Allocation* (ed. T. C. Koopmans), Wiley (1951), 339–47 and 359–73.
62. George B. Dantzig, *Linear Programming and Extensions*, Princeton University Press and the RAND Corporation, 1963, reprinted by Princeton University Press, 1998; "Reminiscences about the origins of linear programming", *Operations Research Letters* 1 (*2*) (1982), 43–48; and "Linear programming", *Operations Research* 50 (*1*) (2002), 42–47.

63. T. E. Harris and F. S. Ross, *Fundamentals of a Method for Evaluating Rail Net Capacities*, Research Memorandum RM-1573, RAND Corporation, 1955.
64. L. R. Ford Jr. and D. R. Fulkerson, "Maximal flow through a network", and "A simple algorithm for finding maximal network flows and an application to the Hitchcock problem", *Canadian Journal of Mathematics* 8 (1956), 399–404, and 9 (1957), 210–18.
65. P. Elias, A. Feinstein, and C. E. Shannon, "A note on the maximum flow through a network", *IRE Transactions on Information Theory* IT-2 (1956), 117–19.
66. L. R. Ford Jr. and D. R. Fulkerson, *Flows in Networks*, Princeton University Press, 1962.
67. Joseph B. Kruskal Jr., "On the shortest spanning subtree of a graph and the traveling salesman problem", *Proceedings of the American Mathematical Society* 7 (1956), 48–50.
68. Jack Edmonds, "Matroids and the greedy algorithm", *Mathematical Programming* 1 (1971), 127–36.
69. R. C. Prim, "Shortest connection networks and some generalizations", *Bell Systems Technical Journal* 36 (November 1957), 1389–401.
70. R. L. Graham and P. Hell, "On the history of the minimum spanning tree problem", *Annals of the History of Computing* 7 (1985), 43–57.
71. Otakar Borůvka, "O jistém problému minimálním" (On a certain minimal problem), *Práce Moravské Přírodovědecké Spolecnosti* (*Brno*) (*Acta Societatis Scientiarum Naturalium Moravicae*) 3 (1926), 37–58, and "Příspěvek k řešení otázky ekonomické stavby elektrovodných sítí" (A contribution to the solution of a problem of economical construction of electrical networks), *Elekronický Obzor* 15 (1926), 153–54. See also Jaroslav Nešetřil, Eva Milková, and Helena Nešetřilová, "Otakar Borůvka on minimum spanning tree problem: Translation of both the 1926 papers, comments, history", *Discrete Mathematics* 231 (2001), 3–36.
72. Vojtěch Jarník, "O jistém problému minimálním" (On a certain minimal problem), *Práce Moravské Přírodovědecké Spolecnosti* (*Brno*) (*Acta Societatis Scientiarum Naturalium Moravicae*) 6 (1930), 57–63.
73. Vojtěch Jarník and Milos Kössler, O minimálních grafech obsahujících n daných bodů (On minimal graphs containing n given points), *Časopis Pro Pěstování Matematiky a Fysiky* 63 (1934), 223–35.
74. G. Dantzig, R. Fulkerson, and S. Johnson, "Solution of a large-scale traveling salesman problem, Paper P-510, RAND Corporation, and *Journal of the Operations Research Society of America* 2 (*4*) (1954), 393–410.
75. David L. Applegate, Robert E. Bixby, Vasek Chvátal, and William J. Cook, *The Traveling Salesman Problem: A Computational Study*, Princeton University Press, 2007; Gregory Gutin and Abraham P. Punnen, *The Traveling Salesman Problem and Its Variations*, Springer, 2007; and William J. Cook, *In Pursuit of the Traveling Salesman: Mathematics at the Limits of Computation*, Princeton University Press, 2012.
76. E. F. Moore, "The shortest path through a maze", *Proceedings of an International Symposium on the Theory of Switching, 2–5 April 1957*, Part II (ed. H. Aiken), *Annals of the Computation Laboratory of Harvard University* 30, Harvard University Press (1959), 285–92.

77. George J. Minty, "A comment on the shortest-route problem", *Operations Research* 5 (1957), 724.
78. E. W. Dijkstra, "A note on two problems in connexion with graphs", *Numerische Mathematik* 1 (1959), 269–71.
79. George B. Dantzig, "Discrete-variable extremum problems", *Operations Research* 5 (*2*) (1957), 266–88.
80. Frank Harary, *Graph Theory*, Addison-Wesley, 1969, p. v.
81. Frank Harary, "On the notion of balance of a signed graph", *Michigan Mathematical Journal* 2 (*2*) (1953–54), 143–46.
82. Frank Harary, Robert Z. Norman, and Dorwin Cartwright, *Structural Models: An Introduction to the Theory of Directed Graphs*, John Wiley & Sons, 1965.
83. Frank Harary, "The number of linear, directed, rooted, and connected graphs", *Transactions of the American Mathematical Society* 78 (1955), 445–63, and "Note on the Pólya and Otter formulas for enumerating trees", *Michigan Mathematical Journal* 3 (1955), 109–12.
84. Frank Harary and Edgar M. Palmer, *Graphical Enumeration*, Academic Press, 1973.
85. F. P. Ramsey, "On a problem in formal logic", *Proceedings of the London Mathematical Society* 30 (1930), 264–86.
86. See Bruce Schechter, "Ronald L. Graham," *Mathematical People: Profiles and Interviews* (ed. D. J. Albers and G. L. Alexanderson), Birkhäuser (Boston) (1985), 110–17, on page 112.
87. R. E. Greenwood and A. M. Gleason, "Combinatorial relations and chromatic graphs", *Canadian Journal of Mathematics* 7 (1955), 1–7.
88. P. Erdős, "Some remarks on the theory of graphs", *Bulletin of the American Mathematical Society* 53 (1947), 292–94.
89. Václav Chvátal and Frank Harary, "Generalized Ramsey theory for graphs, III. Small off-diagonal numbers", *Pacific Journal of Mathematics* 41 (*2*) (1972), 335–45.

CHAPTER 6: THE 1960s AND 1970s

1. Claude Berge, *Théorie des Graphes et ses Applications*, Dunod (Paris), 1958; English translation, *The Theory of Graphs and Its Applications*, Methuen (London), 1962.
2. Oystein Ore, *Theory of Graphs*, Colloquium Publications XXXVIII, American Mathematical Society, 1962.
3. Oystein Ore, *Graphs and Their Uses*, 1963; revised and updated edition by Robin J. Wilson, New Mathematical Library 34, Mathematical Association of America, 1990.
4. Oystein Ore, *The Four-Color Problem*, Academic Press, 1967.
5. G. A. Dirac, "Some theorems on abstract graphs", *Proceedings of the London Mathematical Society* (*3*) 2 (1952), 69–81.
6. Oystein Ore, "Note on Hamilton circuits", *American Mathematical Monthly* 67 (1960), 55.
7. Oystein Ore and Joel Stemple, "On the four color problem", *Notices of the American Mathematical Society* 15 (*1*) (January 1968), 196.

8. O. Ore and J. G. Stemple, "Numerical calculations on the four-color problem", *Journal of Combinatorial Theory* 8 (1970), 65–78.
9. J. Mayer, "Inégalités nouvelles dans le problème des quatre couleurs", *Journal of Combinatorial Theory (B)* 19 (1975), 119–49.
10. Gerhard Ringel, *Map Color Theorem*, Springer-Verlag, 1974.
11. Richard Courant and Herbert Robbins, *What Is Mathematics?*, Oxford University Press (1941), 248.
12. Gerhard Ringel, "Farbensatz für nichtorientierbare Flächen beliebigen Geschlechtes", and "Farbensatz für orientierbare Flächen vom Geschlecht $p>0$", *Journal für die reine und angewandte Mathematik* 190 (1952), 129–47, and 193 (1954), 11–38.
13. Gerhard Ringel, "Bestimmung der Maximalzahl der Nachbargebiete auf nichtorientierbaren Flächen", *Mathematische Annalen* 127 (1954), 181–214.
14. Gerhard Ringel, *Färbungsprobleme auf Flächen und Graphen*, VEB Deutscher Verlag der Wissenschaften (Berlin), 1959.
15. William Gustin, "Orientable embedding of Cayley graphs", *Bulletin of the American Mathematical Society* 69 (*2*) (1963), 272–75.
16. Gerhard Ringel and J.W.T. Youngs, "Solution of the Heawood map-coloring problem", *Proceedings of the National Academy of Sciences of the United States of America* 60 (1968), 438–45.
17. G. Ringel, "J.W.T. Youngs", *Journal of Combinatorial Theory (B)* 13 (1972), 91–93.
18. See *Four Colors Suffice*, on page 126.
19. G.-C. Rota, Letter to the American Mathematical Society formally nominating Ronald L. Graham as AMS president, 1991.
20. Further information about Ron Graham can be found in the Wikipedia article, "Ronald Graham", the St Andrews MacTutor website biography for "Ronald Lewis Graham", and Bruce Schechter, "The peripatetic number juggler", *Discover* (October 1982), 44–47, 50–52, reprinted under the title "Ronald L. Graham," *Mathematical People: Profiles and Interviews* (ed. D. J. Albers and G. L. Alexanderson), Birkhäuser (Boston) (1985), 110–17.
21. See Schechter (note 20), on pages 111, 113.
22. The Stanford University colleague was Persi Diaconis; see Schechter (note 20), on page 111.
23. R. L. Graham and H. O. Pollak, "On the addressing problem for loop switching", *Bell System Technical Journal* 50 (*8*) (1971), 2495–519, and "On embedding graphs in squashed cubes", *Graph Theory and Applications*, Proceedings of the Conference at Western Michigan University, 10–13 May 1972 (ed. Y. Alavi, D. R. Lick, and A. T. White), Lecture Notes in Mathematics 303, Springer-Verlag (1972), 99–110.
24. "Henry Pollak: Interviewed by Donald J. Albers and Michael J. Thibodeaux", in Albers and Alexanderson (eds.) (note 20), on page 243.
25. See Weigan Yan and Yeong-Nan Yeh, "A simple proof of Graham and Pollak's theorem", *Journal of Combinatorial Theory (A)* 113 (2006), 892–93, and D. Zeilberger, "Dodgson's determinant-evaluation rule proved by two-timing men and women", *Electronic Journal of Combinatorics* 4 (*2*) (1997), R22. C. L. Dodgson's original article, "Condensation of determinants", appeared in the *Proceedings of the Royal Society of London* 15 (1866), 150–55.

26. See Martin Aigner and Günter M. Ziegler, *Proofs from THE BOOK*, sixth edition, Springer-Verlag (2018), 79–80.
27. See Ronald L. Graham and Joel H. Spencer, "Ramsey theory", *Scientific American* (July 1990), 112–17, and Ronald L. Graham, Bruce L. Rothschild, and Joel H. Spencer, *Ramsey Theory*, second edition, Volume 20, Wiley Series in Discrete Mathematics and Optimization, 1990.
28. F.R.K. Chung and R. L. Graham, "On multicolor Ramsey numbers for complete bipartite graphs", *Journal of Combinatorial Theory* (*B*) 18 (1975), 164–69.
29. "2003 Steele prizes", *Notices of the American Mathematical Society* 50 (*4*) (April 2003), 465–66.
30. J. von Neumann, "A certain zero-sum two-person game equivalent to the optimal assignment problem", *Contributions to the Theory of Games*, Volume II (ed. H. W. Kuhn and A. W. Tucker), *Annals of Mathematics* Study 28, Princeton University Press, 1963.
31. Alan Cobham, "The intrinsic computational difficulty of functions", *Proceedings of the 1964 International Congress for Logic, Methodology, and Philosophy of Science* (ed. Y. Bar-Hillel), Elsevier/North-Holland (1965), 24–30.
32. Jack Edmonds, "Paths, trees, and flowers", *Canadian Journal of Mathematics* 17 (1965), 449–67.
33. Jack Edmonds, "Maximum matching and a polyhedron with 0–1 vertices", *Journal of Research of the National Bureau of Standards* 69B (1965), 125–30.
34. Jack Edmonds, "Minimum partition of a matroid into independent subsets", *Journal of Research of the National Bureau of Standards* 69B (1965), 67–72.
35. See Edmonds (note 32), on page 451.
36. L. Auslander and S. V. Parter, "On imbedding graphs in the sphere", *Journal of Mathematics and Mechanics* 10 (*3*) (May 1961), 517–23.
37. John Hopcroft and Robert Tarjan, "Efficient planarity testing", *Journal of the Association for Computing Machinery* 21 (*4*) (1974), 549–68.
38. Stephen A. Cook, "The complexity of theorem-proving procedures", *Proceedings of the 3rd Annual ACM Symposium in the Theory of Computing, Shaker Heights, Ohio, 1971*, ACM (1971), 151–58.
39. Richard M. Karp, "Reducibility among combinatorial problems", *Complexity of Computer Computations* (ed. R. E. Miller and J. W. Thatcher), Plenum Press (1972), 85–103, and "On the computational complexity of combinatorial problems", *Networks* 5 (1975), 45–68.
40. Michael R. Garey and David S. Johnson, *Computers and Intractability: A Guide to the Theory of NP-Completeness*, W. H. Freeman, 1979.
41. The material in this section is adapted from *Four Colors Suffice*, Chapters 8, 9, and 10. Other books on the four color problem include Thomas L. Saaty and Paul C. Kainen's *The Four-Color Problem: Assaults and Conquest*, McGraw-Hill, 1977 (reprinted in paperback by Dover, 1986); and Rudolf and Gerda Fritsch's, *The Four-Color Theorem: History, Topological Foundations, and Ideas of Proof*, Springer, 1999. Also of interest, especially on the work of Heesch, Haken, and Appel, and on the philosophical aspects of the solution, is Donald MacKenzie's "Slaying the Kraken: The sociohistory of a mathematical proof", *Social Studies of Science* 29 (*1*) (February 1999), 7–60.

42. Heinrich Heesch, *Untersuchungen zum Vierfarbenproblem*, B. I. Hochschulskripten, 810/810a/810b, Bibliographisches Institut (Mannheim), 1969.
43. For Edward F. Moore's map, see L. Steen, "Solution of the four color problem", *Mathematics Magazine* 49 (*4*) (September 1976), 219–22.
44. Hassler Whitney and W. T. Tutte, "Kempe chains and the four colour problem", *Utilitas Mathematica* 2 (1972), 241–81; reprinted in *Studies in Graph Theory II* (ed. D. R. Fulkerson), Mathematical Association of America (1975), 378–413.
45. The quotations by Haken and Appel appearing here are taken from interviews conducted by Tony Dale on April 16 and June 6, 1994, for Donald MacKenzie's article, "Slaying the Kraken" (see note 41).
46. See note 45.
47. See note 45.
48. K. Appel and W. Haken, "The existence of unavoidable sets of geographically good configurations", *Illinois Journal of Mathematics* 20 (1976), 218–97.
49. See note 45.
50. W. T. Tutte, "Map-coloring problems and chromatic polynomials", *American Scientist* 62 (1974), 702–5.
51. Tutte's verse was quoted by Haken in his interview (see note 45).
52. [Andrew Ortoni, The four color problem], *The Times* (July 23, 1976), London.
53. K. Appel and W. Haken, "Every planar map is four colourable", *Bulletin of the American Mathematical Society* 82 (1976), 711–12.
54. K. Appel and W. Haken, "A proof of the Four Color Theorem", *Discrete Mathematics* 16 (1976), 179–80, and "The solution of the four-color-map problem", *Scientific American* 237 (*4*) (October 1977), 108–21.
55. K. Appel and W. Haken, "Every planar map is four colorable. Part I: Discharging", and K. Appel, W. Haken, and J. Koch, "Every planar map is four colorable. Part II: Reducibility", *Illinois Journal of Mathematics* 21 (1977), 429–90 and 491–567.

AFTERMATH

1. Robert G. Busacker and Thomas L. Saaty, *Finite Graphs and Networks: An Introduction with Applications*, McGraw-Hill, 1965; Mehdi Behzad and Gary Chartrand, *Introduction to the Theory of Graphs*, Allyn and Bacon, 1971; Robin J. Wilson, *Introduction to Graph Theory*, Oliver & Boyd (Edinburgh) and Academic Press (New York), 1972; and J. A. Bondy and U.S.R. Murty, *Graph Theory with Applications*, Macmillan, 1976.
2. John Riordan, *An Introduction to Combinatorial Analysis*, Wiley,1958 (reprinted by Dover Publications, 2003); H. J. Ryser, *Combinatorial Mathematics*, Carus Mathematical Monographs 14, Mathematical Association of America, 1963; Marshall Hall Jr., *Combinatorial Theory*, Blaisdell (Waltham, MA), 1967; C. L. Liu, *Introduction to Combinatorial Mathematics*, McGraw-Hill, 1968; and Ian Anderson, *A First Course in Combinatorial Mathematics*, Clarendon Press (Oxford), 1974.

3. Frank Harary (ed.), *Graph Theory and Theoretical Physics*, Academic Press, 1967, and *A Seminar on Graph Theory*, Holt, Rinehart & Winston, 1967, and Dover Publications, 2015; W. T. Tutte, *Connectivity in Graphs*, University of Toronto Press, 1966; J. W. Moon, *Topics in Tournaments*, Holt, Rinehart & Winston, 1968; J. W. Moon, *Counting Labelled Trees*, Canadian Mathematical Monographs No. 1, 1970; Richard Bellman, Kenneth L. Cooke, and Jo Ann Lockett, *Algorithms, Graphs, and Computers*, Academic Press, 1970; Arthur T. White, *Graphs, Groups and Surfaces*, North-Holland Mathematics Studies 8, 1973; P. Erdős and J. Spencer, *Probabilistic Methods in Combinatorics*, Academic Press, 1974; and Fred S. Roberts, *Discrete Mathematical Models, with Applications to Social, Biological, and Environmental Problems*, Prentice-Hall, 1976.
4. N. L. Biggs, E. K. Lloyd, and R. J. Wilson, *Graph Theory: 1736–1936*, Oxford University Press, 1976; later editions, 1986 and 1998.
5. Lowell W. Beineke and Robin J. Wilson (eds.), *Selected Topics in Graph Theory*, Academic Press, 1978, and *Applications of Graph Theory*, Academic Press, 1979.
6. A. J. Hoffman and R. R. Singleton, "On Moore graphs with diameters 2 and 3", *IBM Journal of Research and Development* 4 (1960), 497–504.
7. S. L. Hakimi, "On the realizability of a set of integers as degrees of the vertices of a graph", *Journal of the Society for Industrial and Applied Mathematics* 10 (1962), 496–506; see also V. Havel, "A remark on the existence of finite graphs" (in Czech), *Časopis pro Pěstování Matematiky* 80 (1955), 477–80.
8. J. W. Moon, "Tournaments with a given automorphism group", *Canadian Journal of Mathematics* 16 (1964), 485–89.
9. W. T. Tutte, "Lectures on matroids" and "Menger's theorem for matroids", *Journal of Research of the National Bureau of Standards* 69B (1965), 1–47 and 49–53.
10. Lowell W. Beineke, "On derived graphs and digraphs", *Beiträge zur Graphentheorie* (ed. H. Sachs, H. J. Voss, and H. Walther), Teubner-Verlag (Leipzig), (1968), 17–23, and "Characterizations of derived graphs", *Journal of Combinatorial Theory* 9 (1970), 129–35.
11. See P. Erdős, "On a problem in graph theory", *Mathematical Gazette* 47 (1963), 220–23, and R. L. Graham and J. H. Spencer, "A constructive solution to a tournament problem", *Canadian Mathematical Bulletin* 14 (*1*) (1971), 45–48.
12. J. A. Bondy, "Pancyclic graphs I", *Journal of Combinatorial Theory* 11 (1971), 80–84, and J. A. Bondy and V. Chvátal, "A method in graph theory", *Discrete Mathematics* 15 (1976), 111–36.
13. Jonathan L. Gross, "Voltage graphs", *Discrete Mathematics* 9 (1974), 239–46.
14. Henry Glover and John Philip Huneke, "The set of irreducible graphs for the projective plane is finite", *Discrete Mathematics* 22 (1978), 243–56.
15. Neil Robertson and P. D. Seymour, "Generalizing Kuratowski's theorem", *Congressus Numerantium* 45 (1984), 129–38, and "Graph Minors XX. Wagner's conjecture", *Journal of Combinatorial Theory* (*B*) 35 (2004), 39–61.
16. Hugo Hadwiger, "Über eine Klassifikation der Streckencomplexe" (On a classification of line complexes), *Vierteljahrsschrift der Naturforschenden Gesellschaft Zürich* 88 (1943), 133–42.
17. Neil Robertson, Paul Seymour, and Robin Thomas, "Hadwiger's conjecture for K_6-free graphs", *Combinatorica* 13 (*3*) (1993), 279–361.

18. Neil Robertson, Daniel P. Sanders, Paul Seymour, and Robin Thomas, "A new proof of the four-colour theorem", *Electronic Research Announcements of the American Mathematical Society* 2 (*1*) (1996), 17–25.
19. In September 2004, Robertson and his team's proof of the four color theorem received a formal machine-checked verification by the French computer scientist George Gonthier; see G. Gonthier, "Formal proof—The four color theorem", *Notices of the American Mathematical Society* 55 (December 2008), 1382–93.
20. C. Berge, "Färbung von Graphen, deren sämtliche bzw. Deren ungerade Kreise starr sind" (Coloring of graphs whose entire (resp. odd) cycles are rigid), *Wissenschaftliche Zeitschrift der Martin-Luther-Universität Halle-Wittenberg* X/1 (1961), 114–15.
21. L. Lovász, "A characterization of perfect graphs", *Journal of Combinatorial Theory* (*B*) 13 (1972), 95–98.
22. M. Chudnovsky, N. Robertson, P. Seymour, and R. Thomas, "The strong perfect graph theorem", *Annals of Mathematics* 164 (2006), 51–229; see also Paul Seymour, "How the proof of the strong perfect graph conjecture was found", *Gazette des Mathématiques* 109 (2006), 69–83.

ACKNOWLEDGMENTS AND PICTURE CREDITS

ACKNOWLEDGMENTS

Many people and institutions have helped in the preparation of this book in a variety of ways. The authors would like to thank the following institutions for support, help, and advice: Bibiothèque centrale du Conservatoire national des arts et métiers (Paris), Clark University, Colorado College, the George Pólya Collection, the János Bolyai Mathematical Society, the Johns Hopkins University, the John Simon Guggenheim Foundation, the Open University, RAND Corporation, Université libre de Bruxelles, the University of Illinois at Urbana–Champaign, the University of Texas at Austin, and the University of West Virginia.

We also wish to thank the following individuals: Ian Anderson, Andrew Appel, June Barrow-Green, André Barnard, Beth Bernstein, Peggy L. Currid, Florence Desnoyers-Robison, Jack Edmonds, Martin Charles Golumbic, Henry Gould, Armgard and Wolfgang Haken, Fred Holroyd, Julie P. Jagerova, Leonard Klosinski, Matjaž Krnc, Sammi Merritt, Péter P. Pálfy, Karen Parshall, Kathleen Quinn, Priscilla Redfield Roe, Cynthia A. Shennette, Jim Stimpert, Sally Whitney Thurston, Bjarne Toft, Clay Ward, Nathaniel J. Weir, Kelvin Yaprianto; and at Princeton University Press, Lauren Bucca, Karen Carter, Eric Crahan, Diana Gillooly, Kristen Hop, Nancy Marcello, and Susannah Shoemaker.

Every effort has been made to trace all copyright holders, but if any have been inadvertently overlooked, the publishers will be pleased to make the necessary arrangements at the earliest opportunity.

PICTURE CREDITS

p. ii: Appel and Haken: courtesy of the Department of Mathematics, University of Illinois; p. 4: The 'Colledges in Cambridge': Wikimedia Commons; p. 5: Yale College: Wikimedia Commons; p. 7: Nassau Hall at Princeton: Wikimedia Commons; p. 9: Great Dome at MIT: Wikimedia Commons; p. 10: Johns Hopkins University: Wikimedia Commons; p. 12: Benjamin Peirce: Wikimedia Commons; p. 15: Eliakim Hastings

Moore: Wikimedia Commons; p. 19: James Joseph Sylvester: Wikimedia Commons; p. 21: *American Journal of Mathematics*, Volume 1 (1878), Title page; p. 23: Edward Frankland and William Kingdon Clifford: Wikimedia Commons; p. 23: Graphic notation: line drawings; p. 25: From J. J. Sylvester, "On an application of the new atomic theory to the graphical representation of the invariants and covariants of binary quantics,—with three appendices", *American Journal of Mathematics* 1 (1878), 64–125, Plate I (available online); p. 29: Arthur Cayley: Wikimedia Commons; p. 31: Alfred Bray Kempe: source unknown; alt: Wikimedia Commons; p. 36: William E. Story: Wikimedia Commons; p. 39: From A. B. Kempe, "On the geographical problem of the four colours", *American Journal of Mathematics* 2 (1879), 193–200, Plate II; p. 45: William Story: courtesy of the Clark University Archives and Special Collections, Robert Hutchings Goddard Library; p. 46: Charles Sanders Peirce: Wikimedia Commons; p. 48: C. S. Peirce: courtesy of the University Archives, Sheridan Libraries, Johns Hopkins University; p. 53: P. G. Tait: Wikimedia Commons; p. 54: Percy John Heawood: courtesy of Robin Wilson; p. 58: Julius Petersen: Wikimedia Commons; p. 59: From J. Petersen, [Sur le théorème de Tait], *L'Intermédiaire des Mathématiciens* 5 (1898), 225–27; p. 61: Lothar Heffter: collection of the Schleswig-Holsteinische Landesbibliothek, Kiel; p. 63: Heinrich Tietze: Wikimedia Commons; p. 67: Hermann Minkowski: Wikimedia Commons; p. 71: Oswald Veblen: Wikimedia Commons; p. 78: George D. Birkhoff: Wikimedia Commons; p. 84: From G. D. Birkhoff, "The reducibility of maps", *American Journal of Mathematics* 35 (1913), 115–28, on page 125; p. 87: Philip Franklin: from the collection of the John Simon Guggenheim Foundation; p. 91: From P. Franklin, "The four color problem", *American Journal of Mathematics* 44 (1922), 225–36, on page 236; p. 93: H. Roy Brahana: courtesy of the University of Illinois Archives; p. 95: J. Howard Redfield: from the collection of the late Mrs. Priscilla Redfield Roe (daughter of J. H. Redfield); p. 98: From J. H. Redfield, "The theory of group-reduced distributions", *American Journal of Mathematics* 49 (1927), 433–55, on page 443; p. 99: From J. H. Redfield (see p. 98), on page 452; p. 99: George Pólya: courtesy of the George Pólya Collection and Leonard Klosinski; p. 101: Clarence N. Reynolds Jr.: courtesy of Henry W. Gould, Emeritus Professor of Mathematics, West Virginia University; p. 106: Dénes König: courtesy of the János Bolyai Mathematical Society; p. 109: König's *Theorie der endlichen und unendlichen Graphen*: two editions; p. 110: Alfred Errera: courtesy of the Université libre de Bruxelles, Archives, Heritage & Rare Books Collection; Title page of thesis, *Du coloriage des cartes et de quelques questions d'analysis situs*, 1921; p. 111: Utilities problem and the graph $K_{3,3}$: H. E. Dudeney, *Amusements in Mathe-*

matics, Thomas Nelson and Sons (1917), Problem 251; line drawing; p. 113: André Sainte-Laguë: courtesy of the Bibiothèque centrale du Conservatoire national des arts et métiers, Paris, ASL I 8; *Les Réseaux* (*ou graphes*), Gauthier-Villars (Paris), 1926; p. 115: Karl Menger: collection of the University of Texas at Austin; p. 117: Kazimierz Kuratowski: courtesy of Robin Wilson; p. 122: Hassler Whitney: with kind permission from Sally Whitney Thurston (daughter of Hassler Whitney); p. 144: Saunders Mac Lane: from the collection of the John Simon Guggenheim Foundation; Wikimedia Commons; p. 151: From *Mathematical Reviews*: Announcement and Volume 1, No. 1, The American Mathematical Society & The Mathematical Association of America (1940), title page; p. 155: W. T. Tutte: courtesy of the University of Waterloo Library, Special Collections & Archives; p. 159: Chromatic polynomials: from G. D. Birkhoff and D. C. Lewis, "Chromatic polynomials", *Transactions of the American Mathematical Society* 60 (1946), 355–451, on page 355; p. 163: Sister Mary Petronia Van Straten: from *Wisconsin Teacher of Mathematics* 69 (*1*) (2017), 5; p. 165: Claude E. Shannon: Wikimedia Commons; p. 168: Bill Tutte at Bletchley Park: with the permission of Jack Copeland; Bill Tutte at Waterloo: courtesy of the University of Waterloo Library, Special Collections & Archives; p. 170: Squared rectangle and associated network: from R. L. Brooks, C.A.B. Smith, A. H. Stone, and W. T. Tutte, "The dissection of rectangles into squares", *Duke Mathematical Journal* 7 (*1*) (1940), 312–40, on page 314; p. 185: Railway network: courtesy of the RAND Corporation; p. 192: Map of state capitals: from R. C. Prim, "Shortest connection networks and some generalizations", *Bell System Technical Journal* 36 (November 1957), 1389–401, on page 1390 (available online); p. 198: Frank Harary in Hungary: copyright granted by Bjarne Toft (Odense, Denmark); p. 204: Oystein Ore: from the collection of the John Simon Guggenheim Foundation; p. 207: Gerhard Ringel and Ted Youngs: courtesy of Robin Wilson; p. 216: Ronald L. Graham: Wikimedia Commons; p. 222: Jack Edmonds: with kind permission by Jack Edmonds; p. 224: Stephen Cook: Wikimedia Commons; p. 227: Wolfgang Haken: courtesy of Armgard and Wolfgang Haken; courtesy of the University of Illinois Archives; p. 235: Kenneth Appel: courtesy of Nathaniel J. Weir (grandson of Kenneth Appel); courtesy of the University of Illinois Archives; p. 238: Computer output from Appel and Haken: courtesy of the Department of Mathematics, University of Illinois; p. 240: Reducible configurations: courtesy of the Department of Mathematics, University of Illinois; p. 243: K. Appel and W. Haken: "Every planar map is four colorable. Part I: Discharging", *Illinois Journal of Mathematics* 21 (1977), 429–90, on page 429.

INDEX